UNIVERSITY OF READING EUROPEAN AND INTERNATIONAL STUDIES

*Published in association with the Graduate School of European and International Studies, University of Reading*

General Editor: Christoph Bluth

This series includes books which discuss some of the major contemporary European and international issues from a comparative perspective. National experiences with a relevance for broader European and international issues are also covered by this series. The collection is interdisciplinary in nature with the aim of bringing together studies that emphasise the role of political, economic, historical and cultural factors in shaping the course of international cooperation and international conflicts, particularly from the point of view of Europe and its relations with the rest of the world. The influence of the processes of European integration (economic, political, cultural) on both Europe and the rest of the world, as well as the impact on Europe of global integration processes and non-European integration schemes, are some of the themes that run through the volumes included in the series.

*Titles include*:

Zygmunt G. Barański and Robert Lumley (*editors*)
CULTURE AND CONFLICT IN POSTWAR ITALY

Jeremy Clegg
MULTINATIONAL ENTERPRISE AND WORLD COMPETITION

K. R. Dark with A. L. Harris
THE NEW WORLD AND THE NEW WORLD ORDER

Anne Deighton (*editor*)
BRITAIN AND THE FIRST COLD WAR

Barry Holden (*editor*)
THE ETHICAL DIMENSIONS OF GLOBAL CHANGE

Yelena Kalyuznova
THE KAZAKSTANI ECONOMY

Athena Leoussi
NATIONALISM AND CLASSICISM

Robert D. Pearce
THE INTERNATIONALIZATION OF RESEARCH AND DEVELOPMENT BY MULTINATIONAL ENTERPRISES

Robert D. Pearce and Satwinder Singh
GLOBALIZING RESEARCH AND DEVELOPMENT

Stelios Stavridis, Theodore Couloumbis, Thanos Veremis and Neville Waites (*editors*)
THE FOREIGN POLICIES OF THE EUROPEAN UNION'S MEDITERRANEAN STATES AND APPLICANT COUNTRIES IN THE 1990s

**University of Reading European and International Studies**
**Series Standing Order ISBN 0–333–71510–1**
(*outside North America only*)

You can receive future titles in this series as they are published by placing a standing order. Please contact your bookseller or, in case of difficulty, write to us at the address below with your name and address, the title of the series and the ISBN quoted above.

Customer Services Department, Macmillan Distribution Ltd
Houndmills, Basingstoke, Hampshire RG21 6XS, England

# The Foreign Policies of the European Union's Mediterranean States and Applicant Countries in the 1990s

Edited by

Stelios Stavridis
*Director*
*Centre for Euro-Mediterranean Studies*
*University of Reading*

Theodore Couloumbis
*Professor of International Relations*
*University of Athens*
*Greece*

Thanos Veremis
*Professor of Political History*
*University of Athens*
*Greece*

and

Neville Waites
*Deputy Director*
*Centre for Euro-Mediterranean Studies*
*University of Reading*



in association with
GRADUATE SCHOOL OF EUROPEAN AND INTERNATIONAL STUDIES
UNIVERSITY OF READING

First published in Great Britain 1999 by
**MACMILLAN PRESS LTD**
Houndmills, Basingstoke, Hampshire RG21 6XS and London
Companies and representatives throughout the world

A catalogue record for this book is available from the British Library.

ISBN 0–333–75100–0

First published in the United States of America 1999 by
**ST. MARTIN'S PRESS, INC.,**
Scholarly and Reference Division,
175 Fifth Avenue, New York, N.Y. 10010



ISBN 0–312–21916–4

Library of Congress Cataloging-in-Publication Data
The foreign policies of the European Union's Mediterranean states and applicant countries in the 1990s / edited by Stelios Stavridis . . . [et al.].
p. cm. — (University of Reading European and international studies)
Includes bibliographical references and index.
ISBN 0–312–21916–4 (cloth)
1. European Union—Mediterranean Region. 2. Mediterranean Region–
–Foreign relations—1945– I. Stavridis, Stelios. II. Series.
D1065.M628F67 1998
327'.09182'209045—dc21 98–37120
CIP



This book is produced by the Centre for Euro-Mediterranean Studies, University of Reading, and ELIAMEP, University of Athens. It was supported by the Eleni Nakou Foundation but the latter does not bear any responsibility for views expressed by the authors of the book.



This book is printed on paper suitable for recycling and made from fully managed and sustained forest sources.

10 9 8 7 6 5 4 3 2 1
08 07 06 05 04 03 02 01 00 99

Printed and bound in Great Britain by Antony Rowe Ltd, Chippenham, Wiltshire

# Contents

# List of Contributors

| | |
|---|---|
| **Roberto Aliboni** | Istituto Affari Internazionali, Rome |
| **Theodore Couloumbis** | The University of Athens and ELIAMEP, Athens |
| **Carlos Echeverría Jesús** | WEU Institute for Security Studies, Paris |
| **Fernanda Faria** | Instituto de Estudos Estratégicos e Internacionais, Lisbon |
| **P. C. Ioakimidis** | University of Athens and ELIAMEP, Athens |
| **Kemal Kirisçi** | Bogaziçi University, Istanbul |
| **Laurent Meyrede** | Fondation Méditerranéenne d'Etudes Stratégiques, Paris |
| **Yannis Nicolau** | ELIAMEP, Athens |
| **Roderick Pace** | University of Malta |
| **Joel Peters** | The University of Reading |
| **Stelios Stavridis** | The University of Reading |
| **Charalambos Tsardanidis** | Institute of International Economic Relations, Athens |
| **Thanos Veremis** | University of Athens and ELIAMEP, Athens |
| **Neville Waites** | The University of Reading |

# Acknowledgements

The editors, from the Centre for Euro-Mediterranean Studies at the University of Reading and the Hellenic Foundation for European and Foreign Policy (ELIAMEP), Athens, express appreciation for a financial award from the Eleni Nakou Foundation, London, which made this research project and publication possible.

# Introduction: The Mediterranean in Perspective

Theodore Couloumbis
and Thanos Veremis

Raymond Aron in his classic treatise on international relations[1] divided his analytical approach into four categories which he called history, sociology, theory and praxeology.[2] In presenting our introductory thoughts on Europe and the Mediterranean, we shall employ our version of the thoughtful Aronian taxonomy, raising some difficult questions in the case of 'history' and 'sociology', juxtaposing alternative conceptual paradigms in the case of 'theory', and arriving at some policy responses and recommendations in the case of 'praxeology'.

## HISTORY

The Mediterranean – a remarkable body of water – gave birth to some of the world's great civilisations, religions and cultures. Over the centuries, before and after the birth of Christ, the littoral regions of the dark wine sea experienced waves of migrations, the formation of city states, and the rise of great empires such as the Hellenistic, the Roman, the Byzantine, the Arabic, the Ottoman and the British.

Yet, whereas the Hellenic–Judaic tradition conquered the minds and hearts of the Europeans with little resistance, Islam failed to make significant inroads in the West. The Ottomans left their religious heritage in Bosnia and Albania but the Arabs that preceded them facilitated the transmission of Aristotelian thought into the Europe of the tenth century. The subsequent blooming of the Renaissance was assisted by the Byzantine transfusion of classical Greek philosophy and Platonic thought that questioned the established Aristotelian wisdom.

Europe's emergence from the waters of the miraculous sea as a child of the Hellenic–Judaic tradition was followed by the gradual extinction of the same tradition in the eastern and southern shores of the Mediterranean. 'This epochal event, to a great extent established the cultural division between East and West, North and South in the Mediterranean.'[3] The division persists to this day as an east–west and north–south cultural and economic cleavage. The Mediterranean in fact was only a cultural unit in the Hellenistic and Roman periods. After the decline of Byzantium the region became a frontier of hostile civilisations or, at best, a junction of diversity expressed in terms of economic development, degrees of secular modernity and the nature of political authority. As Vatikiotis points out, on the European side of the Mediterranean there is a concept of natural law which does not exist on the other side of the sea. 'Consequently, such matters as individual rights deriving from this higher law, to protect the individual from the excesses of the state, are difficult to formulate and maintain.'[4]

When the empires that had politically unified parts of the Mediterranean came to an end, the sea and the adjacent lands became the battlefield of disputes between great powers. Most of them – Britain, Russia, Austria, Germany, the United States and the Soviet Union – determined the fate of the region without even being part of it. Many wars of these powers were waged and fought by proxy, through Balkan or Middle East surrogates. Little concern for the damage inflicted upon the region and its people was displayed by the great powers as long as their strategic aims were realised.

The strategic importance of the Mediterranean was eclipsed at least twice in history. Once by naval technology, which shifted the traffic of sea commerce to the Atlantic, then during the Cold War period, when the central front of the European continent attracted most NATO attention. The Western alliance turned to its southern flank when the Middle East crises erupted and their impact on the price of oil became apparent. In our post-Cold War era the strategic importance of the Mediterranean is once more being upgraded. The current setback in the Arab–Israeli peace process may be a temporary one, but the Persian Gulf will remain a trouble spot, the Greek–Turkish problem will still beg a solution, and the Caspian Sea oil will certainly draw international attention early in the coming century.

The tug of war between Russia and Turkey over the conduit of Caspian and Kazakh oil reflects old regional tensions. Turkey and the western companies prefer a pipeline that will cross from Baku-Azerbaijan into Georgia and from there to the Yamourtalik harbour of Hatai. Russia is determined to secure a route for a new pipeline, half-way to the Russian Black Sea harbour of Novorosisk. The battle for positions in the Caspian pipeline competition threatens to pit the interested western companies and Turkey against Russia. Could a Crimean war scenario recur in the twenty-first century that would isolate Russia from Europe once more? The answer to the possibility of such a repetition of the past lies in other historical instances when Russia was drawn into the European nexus of alliances and the Concert of Europe. The present NATO Partnership for Peace, which is something of an antechamber to the community of western European states, could act as the beginning of a reintegration of Russia into the European family.

## SOCIOLOGY

In the evolving post-Cold War architecture of international politics the European Union, together with North America, Japan and the remaining advanced states belonging to the OECD, comprise a global platform of political stability and economic interdependence. Unfortunately, vast regions of what we used to call the Third World as well as parts of the post-Communist Balkans and the former Soviet Union, are still in a state of real or potential fluidity and instability.

The Mediterranean region offers a most dramatic illustration of the structural problems that beset the global divide and which can be summarised in one word: inequality. For example, the collective GDP of the EU Mediterranean states in the North is eleven times greater than that of its southern littoral counterparts.[5]

With the passage of time, this gap in economic levels of wealth and development is growing rather than shrinking. The countries of the North tend to be importers of raw materials (especially oil) and exporters of finished goods. The southern Mediterranean countries are raw material and agricultural product exporters and they import (maintaining a sizable trade imbalance) finished goods. In the EU-North the economies are capital-intensive and the cost of labour is high while in the South the economies are labour-intensive and labour is cheap. Both in the North and the South there are sizable rates of unemployment, but the EU Mediterranean states afford safety net mechanisms and social welfare institutions that tend to cushion the shocks, tempering attendant challenges to political stability.

This unpleasant (especially for the South) situation is further exacerbated if we take into consideration demographic trends in the EU-North and the South respectively. The northern littoral states (with the exception of Turkey) are hovering near zero or even negative population growth rates. In the South and the Northeast (with the exception of Israel) the countries are experiencing a virtual population explosion. By the year 2015 the total population of the fifteen EU members will increase by only 13 million, while the population of the non-EU Mediterranean countries will number an additional 170 million persons. By the year 2015 the combined population of Morocco, Algeria, Egypt and Turkey will amount to 270 million souls – mostly young and unemployed – who will throng in the unfriendly slums of major urban centres.[6] This is indeed an explosive mix and it could easily result in civil conflict, revolutions, military intervention in politics and wars that could produce major population shifts (political and economic refugees, and illegal immigration), which in turn could trigger social, political and economic upheavals among the so-called privileged regions of the North.

As Tables 1 and 2 unequivocally demonstrate, there is relative abundance in the North and abject poverty in the South. Whether we are talking about nutrition, housing, health care, literacy or environmental protection, the gap is huge and it is growing.[7]

On another very sensitive statistical indicator – income distribution – one could talk about societies in which two out of three people are in the comfortable (middle-class) category in the EU-North and one out of ten in the less egalitarian South,

*Table 1* Balance sheet of human development – Arab states

| PROGRESS | DEPRIVATION |
| --- | --- |
| *HEALTH* | |
| In 12 of the 19 countries in the region life expectancy is more than 65 years, compared with an average of 45 years in 1960. | Only half the people in rural areas have access to safe water, and only a third have access to basic sanitation. |
| *EDUCATION* | |
| Over two decades the adult literacy rate almost doubled: from 30 per cent in 1970 to 54 per cent to 1992. | About 60 million of the region's 240 million people are illiterate. |
| Between 1960 and 1990 the primary enrolment ratio more than doubled, from 38 to 77 per cent, and the secondary enrolment ratio nearly tripled, from 18 to 51 per cent. | Nine million children do not attend primary school, and 15 million do not attend secondary school. |
| *INCOME AND POVERTY* | |
| Between 1960 and 1993 real per capita income grew by nearly 3 per cent a year.<br>In 1980–90 the agricultural growth rate was nearly 5 per cent a year – the highest among the developing regions. | About 73 million people still live below the poverty line, and more than 10 million are underfed. |
| *WOMEN* | |
| At the secondary level the number of girls per 100 boys rose from 47 to 77 between 1970 and 1990, and at the tertiary level it rose from 34 to 65.<br>Of women enrolled at the tertiary level, about 30 per cent are in natural or applied sciences. | Only 25 per cent of Arab women participate in the formal labour force, compared with 39 per cent in the developing countries as a group.<br>Women hold only 4 per cent of parliamentary seats, well below the 10 per cent average for the developing world. |

| *CHILDREN* | |
|---|---|
| Between 1960 and 1993 the infant mortality rate declined by more than three-fifths: from 167 per thousand live births to 66.<br>More than three-quarters of one-year-olds are immunized. | At 73 per thousand live births, the under-five mortality rate is still more than four times as high as that in industrial countries. |
| *ENVIRONMENT* | |
| Between 1965 and 1991 energy use per $100 of GDP declined by two-thirds, from 228 kilograms of oil equivalent to 76. | With less than 1,000 cubic metres of water per capita available each year, about 55 per cent of the population suffer from serious water scarcity. |
| *POLITICS AND CONFLICTS* | |
| During the past two decades six countries have held multiparty parliamentary elections, and since 1990 there have been 18 general elections.<br>In 1994, 250,000 refugees in the region returned to their country of origin. | At the end of 1994 more than one million people were refugees. |

*Source*: *Human Development Report 1996*, UNDP, New York and Oxford: Oxford University Press, 1996.

where overwhelming majorities live well below the level of poverty feeding the fires of revolutionary movements (mostly of the anti-Western fundamentalist variety) that could be likened to a time bomb with a short fuse.

In the cultural area, there is a juxtaposition of social, religious and attitudinal differences. One is tempted to talk of a 'Christian Club' when referring to the Mediterranean countries that are currently members of the EU. The rest of the Mediterranean is predominantly Islamic, with the exception again of Israel. And here lies a danger of correlating Islam with authoritarian political systems and Christianity with pluralistic democratic ones.[8] Suffice it to say, at this

*Table 2* Human Development Index*

| *HDI rank* | *Sequence of rankings 1–175* | *Life expectancy at birth (years) 1993* | *Adult literacy rate (%) 1993* | *Combined first -, second- and third-level gross enrolment ratio (%) 1993* | *Real GDP per capita (PPP$) 1993* |
|---|---|---|---|---|---|
| France | 7th | 77.0 | 99.0 | 88 | 19,140 |
| Spain | 10th | 77.7 | 98.0 | 87 | 13,660 |
| Italy | 20th | 77.6 | 97.4 | 70 | 18,160 |
| Greece | 21st | 77.7 | 93.8 | 78 | 8,950 |
| Cyprus | 23d | 77.1 | 94.0 | 76 | 14,060 |
| Israel | 24th | 76.6 | 95.0 | 76 | 15,130 |
| Malta | 28th | 76.2 | 87.0 | 76 | 11,570 |
| Portugal | 35th | 74.7 | 86.2 | 79 | 10,720 |
| Libyan Arab Jamahiriya | 59th | 63.4 | 73.7 | 88 | 6,125 |
| Bulgaria | 62nd | 71.2 | 93.0 | 65 | 4,320 |
| Algeria | 69th | 67.3 | 58.8 | 66 | 5,570 |
| Jordan | 70th | 68.1 | 84.8 | 66 | 4,380 |
| Tunisia | 78th | 68.0 | 64.1 | 66 | 4,950 |
| Turkey | 84th | 66.7 | 81.1 | 62 | 4,210 |
| Syrian Arab Rep. | 92nd | 67.3 | 68.7 | 65 | 4,196 |
| Lebanon | 97th | 68.7 | 91.7 | 74 | 2,500 |
| Albania | 104th | 72.0 | 85.0 | 59 | 2,200 |
| Egypt | 106th | 63.9 | 49.8 | 69 | 3,800 |
| Morocco | 123d | 63.6 | 41.7 | 44 | 3,270 |

* The HDI is based on three indicators: *longevity*, as measured by life expectancy at birth; *educational attainment*, as measured by a combination of adult literacy (two-thirds weight) and combined primary, secondary and tertiary enrolment ratios (one-third weight); and *standard of living* as measured by real GDP per capita.

*Source*: *Human Development Report 1996*, UNDP, New York and Oxford: Oxford University Press, 1996, pp. 135–6.

point, that in using correlation as a substitute for causation, one could argue that Christian Catholicism and Christian Orthodoxy have correlated positively in the recent past with the incidence of military/authoritarian regimes in Southern Europe (until the mid 1970s) and in much of Latin America (well into the 1980s). Yet careful quantitative studies[9] have posited that authoritarian systems and military dictatorships have been more likely to flourish in countries with low levels of economic development and minuscule middle classes (that cannot sustain civil societies) and that democracy can more easily be consolidated in industrial and post-industrial societies with sizable middle classes.

The central question, therefore, that the picture of great structural inequality poses, is whether the Mediterranean region can gradually evolve into a cohesive space of security and cooperation rather than a region of growing suspicion and North –South mutual alienation. We will address this and related questions, in the next two sections of this chapter.

## THEORY

Focusing on the Mediterranean region in the period since World War II, we can employ three distinct conceptual paradigms to help us enhance our understanding of past patterns of behaviour and to project developments in the near and the medium term. The first and second paradigms (or conceptual frameworks) are offshoots of traditional Realism. They both emphasise the centrality of state actors but tend to focus on *conflict* (the first paradigm) and on *consensus* (the second paradigm) among the world's great powers.[10] We can also call them 'conflict' and 'concert' global systems respectively. The third paradigm asserts that non-state actors, in addition to nation states, are playing a growing role in international affairs and that the global system, once described as bipolar or multipolar, depending on the number of relevant great powers, is now becoming a system of complex-interdependence.[11]

Clearly, relations in the Mediterranean region during the Cold War years (1947–89) could be best understood as being derivative of three major sub-patterns of the conflict paradigm. The first sub-pattern was reflected in the intense bipolar confrontation between the forces of Soviet-controlled communism versus the Western coalition of democracies led by the United States. In such a setting the northern littoral countries of the Mediterranean (from Portugal to Turkey) joined the Western strategic sphere (Spain joining NATO only after the restoration of democracy in the post-Franco era). The southern littoral states, in turn, joined the ranks of the non-aligned, post-colonial states. They reflected nonetheless security and political connections and/or dependencies with one Cold War superpower or the other. It is not, for example, coincidental that, during the Cold War, Morocco procured its weapons mainly from the West, Algeria from the East, Tunisia from the West, Libya (after King Idris) from the East, Egypt (after Nasser) from the West, the Palestinians from the East, Jordan from the West, Syria from the East and so forth.

The second sub-pattern of the conflict paradigm pitted ex-colonial powers of the Mediterranean North (Spain, France and Italy) against post-colonial new states of the South. The Algerian war in the 1950s dominated the anti-colonial movement which led to widespread and voluntary decolonisation in the 1960s. Once the European powers, however, digested the imperatives of decolonisation, patterns of cooperation between the post-colonial governments and the old metropolitan centres were once again developed.

The third sub-pattern of conflict has centred around the protracted confrontation between the state of Israel and its Arab neighbours (together with their supporters). This protracted conflict was heavily exacerbated, as well as modified, by the impact of the Cold War. Israel enjoyed the unqualified support of the United States while the Soviet Union tilted in favour of Syria and the Palestinians. The Arab governments, however, became gradually stratified between so-called moderates (supported by the West) and so-called radicals (supported by the former Soviet Union).

The end of the Cold War has ushered in a new period that fits the explanatory framework of the concert/consensus paradigm of Realism. With the breakdown of Soviet controlled communism, the dissolution of the Warsaw Pact, the disaggregation of the former Soviet Union and the reunification of Germany, the central premises of global politics have been radically altered. The cessation of the Cold War and the ideological and political convergence of the United States and Russia (the nuclear power that succeeded the Soviet Union) have given an opportunity to the UN Security Council to reactivate itself and to authorise collective action in critical areas such as Iraq/Kuwait and Bosnia. One could even argue that the peace process between the Israeli government and the PLO was a product, among other factors, of the realisation by the Palestinians that support emanating from the former Soviet Union would no longer be available.

Paradoxically, the end of the Cold War, and of the mutually neutralising system of the nuclear balance of terror, has made it possible for new actors and forces – reflecting nationalist, ethnic and religious fundamentalist values – to surface and to plunge regions of former Yugoslavia and the former Soviet Union into 'clash of civilisations' type wars.[12] In an ironic twist of fate the resumption of cooperative relations among the world's superpowers has increased the incidence of low-grade (local war) conflicts in the so-called periphery of our planet.

The major question in the Mediterranean region in the decades to come will be whether conflict-prone zones such as the Middle East, the Balkans and North Africa will manage to integrate themselves into an emerging zone of stability (the centre) or cling atavistically to patterns of local and externally manipulable conflict (the periphery). It is clearly in the interest of all the Mediterranean regional states to adopt policies and to engage in programmes designed to promote economic development and political convergence in a climate of genuine interdependence.

As we are entering the gates of the twenty-first century (and the third millennium AD), the central question, in theory as well as in practice, is whether the Mediterranean region, despite its structural asymmetries discussed in detail above, will manage to fit into a functionalist paradigm which permitted Western Europe, after two bloody world wars, to move forward toward economic and political integration employing the geoeconomic premises of Jean Monnet and abandoning the military power considerations of Clausewitzian geopolitics. Monnet has offered us a paradigm that defines 'national interests' in collective/regional terms, that abandons territorial logic (irredentism, revisionism, expansionism, revanchism), and which presupposes the willingness of cooperating governments to accept one another's legitimacy and reduce emphasis on sovereignty and national independence in favour of economic interdependence and respect for the values of democracy, pluralism and human rights.

If we accept the premise that the existence of democratic states with civil (pluralist) societies is a central prerequisite for the development of systems of regional integration and interdependence, then we must pose a set of difficult and even uncomfortable questions. Are the low levels of economic development and the inequality in incomes and privileges (in education, health care *et al.*), which are typical characteristics of most non-EU Mediterranean states, likely to thwart efforts to establish legitimate and viable democratic institutions? Employing for a moment Huntingtonian logic[13] could we argue that cultural/historical traditions such as the Islamic, the Catholic and the Slavic-Orthodox are incompatible with the values and institutions of democracy and human rights? Would a system of asymmetric-interdependence (a polite rendering of the concept of dependency) gravitate into new and veiled forms of neocolonialism and exploitation of the poor South by the rich North?

We will return to some of these important questions in the last section of our introduction (on praxeology). We should hasten to state, at this point, that we share the view that consolidated democracy is a prerequisite for EU-type integration. But we discard the assumption that pluralist democracy is only compatible with Judeo-Christian values (e.g. the post-World War II record of India and Japan), and we believe that the old-style 'fortress Europe' logic, perpetuating political and cultural hierarchies, is most likely to explode in the face of the privileged and advanced nations of our planet.

## PRAXEOLOGY

As the relevant chapters in this volume will suggest, France together with Italy, Portugal and Spain have sought to coordinate their policies vis-à-vis the western Mediterranean. Their major preoccupation – especially in France and Spain – has been instability, internal conflict and the rise of anti-Western Islamic fundamentalism (as manifested in Algeria). Their collective fear is that revolutionary conditions and interstate conflict in the Maghreb will unleash tidal waves of refugees flooding the northern shores of the Mediterranean to destabilise social and political balances, leading to increased unemployment as well as to the rise of racist, exclusivist and chauvinistic political forces throughout Western Europe.

In the EU, Portugal, Greece, the South of Italy and, to a certain extent, Spain share some of the features of the least developed regions of the EU and for that reason they can act as bridges between the rich North and the poor South of the ancient sea. On Mediterranean questions, their priorities have been determined mainly by their location and other specific concerns. The proximity of Portugal, Spain and Italy to North Africa and the common borders of Greece and Italy with the former communist Balkans, help explain each country's regional vocation.[14] Yet the constant challenges in the region have endowed these states with a common task. Besides the traditional threats to security posed by regional conflict and the danger of weapons proliferation in the Middle East, new items have been added to the security agenda: the collapse of the economies, the denial of human rights, the rise of terrorism, narcotics trafficking, the persecution of minorities and the degradation of the environment.[15]

Although traditional security issues continue to elude institutionalisation between the EU and Mediterranean non-member states, the new security agenda may offer potential for much closer co-operation and policy co-ordination between the two sides. Migration constitutes the gravest challenge for EU member by non-member states. Linked to the migration issue are the questions of integration, assimilation and citizenship of migrants within the EU, and the relevant issue of Western preoccupation with Islamic fundamentalism. Developments in Algeria have generalised the depiction of the danger in the Western press.[16]

The turbulent Middle East, which has for decades drawn international attention to the Mediterranean, is being tempered by the Arab–Israeli peace process. Yet, the proliferation of weapons of mass destruction and of their means of delivery is nowhere more acute than in this region, where several states possess such items. Chemical weapons and ballistic missiles have been used in conflict and the collapse of the Soviet Union has added to the availability of know-how for their construction.[17] Tensions in the Arab world could re-emerge over the rise of religious fundamentalism in Egypt, a collapse of the Arab–Israeli peace process, support for international terrorism or even over strict Western migration policies. Although the capabilities of states that possess weapons of mass destruction are limited, the stake of the West in availability of Middle East oil is large and will grow with the emergence of China and India as significant energy consumers.[18] Pacifying the region and maintaining cooperation cannot be limited to non-proliferation.

Iran's exclusion from Gulf politics after the Gulf War and US military presence in the region have predictably turned Iranian efforts northward. Although it will be long before Russia returns as a major player in Middle East affairs, it is certain that she is already a major protagonist in the Caspian region. Relations between Russia and Iran are vital in the military sector. Russia maintains that its cooperation with Iran is based on principles determined by the talks on conventional arms sales, initiated by the five permanent members of the UN Security Council, and that weapons will only be sold for self-defence. Arms exports to Iran are important to both sides because Russia needs the cash and Iran the arms. The continuation, however, of arms exports, and more so the transfer of nuclear technology, will no doubt have an adverse effect on Russo–American relations. On the other hand Russia's role in the Arab–Israeli peace process and its aspiration to play a greater role in the future might in time have a moderating impact on Iranian policy.[19]

Post-Soviet Russia has a stake in the improvement of the intractable problems between Iran and the United States. Such an improvement would facilitate Russia's role in Transcaucasia, would reduce Turkey's importance in Western calculations, and would ultimately increase Iran's role in the Middle East and Russia's input in the peace process.

The factor that has brought Iran and Turkey together in the past was a common fear of Soviet power. In almost every other respect, relations between the two states have been a zero sum game vis-à-vis the West. In the oil crisis of 1973 Iran's position was enhanced at the expense of Turkey's strategic importance. The Khomeini revolution and Iran's subsequent estrangement from the West constituted the greatest strategic benefit that Turkey has gained in the region. Turkey has since assumed most of Iran's Western role, has tried to isolate it from its regional links in the pipeline competition, and has widely advertised its own strategic importance as a buffer against the spreading of Islamic fundamentalism.[20] Furthermore, Turkey's continuing interest in the Caucasus and Central Asia is creating the structural foundations for a future rift in Russo–Turkish relations.[21]

The route of pipelines that will carry oil from the Caspian Sea to the West has been a major instrument of policy and a decision to a large degree dependent on political considerations. Russia, as the major military actor in the region, lacking resources to make the necessary investment for such a project, relies mainly on its mediation in the Armenian–Azeri dispute in order to secure that the pipeline will end in the Russian harbour of Novorosisk (from there oil will be shipped through the Black Sea). Turkey and the United States have convinced the oil companies to overrule the Iranian conduit and support a route through Turkey.[22]

In spite of the economic strength of its own Western partners, Turkey has taken up many formidable foes simultaneously: Russia, Iran, Armenia, Syria and the Kurds. Neither Britain nor even the US can project their military might in this remote region. Besides playing their hand in the peace negotiations between Armenians and Azeris, the Russians use their monopoly on existing pipelines to dissuade Azerbaijan and Central Asian states from considering new pipelines that will bypass Russia. Furthermore, Russia, in the October 1993 meeting of the littoral states on the Caspian Sea, pushed to define the latter as a sea rather than a lake. That definition signifies Baku's loss of all major off-shore oilfields because waters beyond the 12-mile limit will be declared neutral, placing the sea under Russia's naval influence.[23]

Italy's and Greece's Balkan interests are, given their proximity, more vital than those of other EU members. Committed to the territorial status quo, both initially opposed instant and unqualified recognition of secessionist Yugoslav states and counselled caution to their EU partners before an alternative state arrangement was explored. The collapse of Yugoslavia in AFSOUTH's region of responsibility has produced the most conspicuous 'failed state' in Europe.[24]

The Bosnian settlement of 21 November 1995 in Dayton, Ohio, is certainly a hopeful development in the troubled region, but peace is still not consolidated. The lifting of the international embargo and the reconstruction of the former Yugoslavia has opened a wide field of activity to the private sector. The Greek government has much to offer in facilitating the exodus of Serbia from its present isolation vis-à-vis the international community and could also act as an interlocutor in future negotiations for an association agreement between the European Union and candidates from the region of former Yugoslavia.

Lebanon, Sudan and Algeria might be added to the Mediterranean list of 'failed states' and the outcome of their 'inability to bring into appropriate balance the goals of sustainable state power, wealth and welfare'.[25] Failed states lead to geopolitical vacuums and therefore conflict between suitors that compete for their domination. Of equal concern for regional security are the 'troubled states' in the Mediterranean: Iran, Iraq, Turkey, Egypt and Tunisia. The frailties of their domestic systems, their inability to resolve ethnic and ideological cleavages without violence, and the isolation of some from the outside world, might be the ingredients of 'troubled states'. Yet neither 'failed' nor 'troubled states' are a new feature in the South of the Mediterranean basin. While the North was structurally bipolar and stable in its adversarial relations between NATO and the Warsaw Pact, the South remained unstable and conflict-prone. Furthermore, throughout the Cold War period the UN was veto-blocked, limiting itself to marginal peacekeeping activities in the Middle East and Africa.

The Maghreb region is undergoing the difficulties of transition caused by high birth-rates, unemployment, debts and Islamic radical movements. The legitimacy of ruling elites has been undermined by their failure to deliver modernisation accompanied by economic growth and development. Although Islam, as a manifestation of discontent, is popular, it is questionable if it can succeed where traditional leaderships failed. As it has been aptly suggested 'European states are seeking greater engagement in the region but are disinclined to bolster existing regimes faced with opposition groups, and recognise that future progress will occur primarily through the efforts of local peoples and elites.'[26]

A serious credibility gap hinders any solution of the Greek–Turkish dispute. The consideration of Cyprus as a full member of the EU however offers a unique opportunity, not only for the resolution of the intractable Cyprus problem, but also for the creation of a significant building bloc in Greek–Turkish relations. Should Turkey advise the Turkish Cypriot leadership not to resist an agreement between the two communities that would facilitate the accession of the island to the EU, Greece could champion Turkey's future entry. This would constitute the most important confidence-building measure between the two states. There is little doubt that if relations deteriorate to the point of armed confrontation, not only the two states, but the entire region will suffer a great setback. A war between Greece and Turkey would destroy Greece's prospects of convergence with the EMU requirements and would exclude Turkey from the European family. The eastern Mediterranean would become a shore of chronic hostility.

Will the post-Cold War era witness a new world order of generalised stability based on respect for the territorial integrity of states, enhancement and consolidation of democratic institutions, the protection of human rights and the institutionalisation of structures for the peaceful settlement of intrastate disputes, or are we moving toward a period of atavistic disorder, ethnic conflict, disorientation and escalating economic protectionism, all resulting in a generalised instability? Will the 'limited wars' which have been taking place in the troubled southern regions of the planet also become a phenomenon of the North as we have been witnessing in former Yugoslavia? To avoid such a possibility the new architecture of global security should be based on an implicit consensus on fundamental premises shared by the world's major centres of power. This arrangement also requires participation of a considerable number of smaller states. A series of interlocking international institutions of economic and political cooperation with the United Nations also need to be developed.

It would be a serious mistake for European states to engage in a debate juxtaposing the merits and demerits of a 'Europeanist' versus an 'Atlanticist' choice. It would be equally mistaken for North Americans to view progress in the European Union as being at odds with the interests and objectives of a post-Cold War NATO. The crucial challenge facing the world today rests on the need to establish a set of complementary and overlapping security structures in areas of actual and potential conflict.[27]

## A PARTING THOUGHT

We have raised in this introductory chapter a number of dilemmas facing the Mediterranean region, its people and their governments from the perspectives of history, sociology, theory and praxeology. The problems are many and they are described variously, depending on the location, capabilities, objectives and interests of the countries surveyed.

The fundamental question which we must answer from a praxeological (normative) perspective is whether the differences and distances in the Mediterranean region (North, South, East and West) make it likely, difficult or impossible to generate lasting international and transnational regimes of peaceful and cooperative coexistence. In short, the question is whether the Mediterranean sea will function as a bridge or a barrier.

From the perspective of each of the countries whose foreign policies are surveyed in this volume, there is every incentive to widen and solidify patterns of peace and interdependence in the wider Mediterranean space. If the EU member states fail to contribute substantively and in a timely fashion to the process of economic development of the non-EU Mediterranean states, then they will be undermining the very foundations of peace, stability and prosperity in their own societies. Economic development in the South will contribute to political stability, lower unemployment and birth-rates, and reduce the tension in the current zones of high conflict quotients.

The EU strategy for security and cooperation in the Mediterranean must be multi-dimensional. In the cultural dimension it must cultivate a spirit of mutual understanding and appreciation that will transform the Mediterranean into a region of peaceful and cooperative projects that take advantage of the long and rich religious and secular traditions of Christianity, Islam and Judaism.

In the economic dimension, cooperation should not be considered 'philanthropy' by the privileged of the North. The European Union's investments in enterprises that are labour-intensive will invigorate the economies of the South and will reduce unemployment and resultant pressures for legal and illegal immigration. The removal by the EU of protectionist measures targeting the products of the Mediterranean South will permit the growth of exports from South to North and simultaneously will reduce governmental methods of the past that have been perpetuating state monopolies and inefficient public-owned enterprises in the South. Simultaneously, MEDA and other regional credit, investment and cooperation programmes will gradually reverse current trends in the so-called braindrain and will permit the citizens of the developing states in the Mediterranean to resist the siren songs of populism, nationalist fanaticism and fundamentalist exclusivism.

In the political dimension, the institutionalisation of bilateral, sub-regional and multilateral mechanisms for tension reduction, conflict resolution, and (in cases of conflict) peace-keeping, peace-building and humanitarian operations should be at the top of the EU's priorities list. But the EU should resist the temptation of becoming a *participant* rather than an *intermediary* in potential conflict situations.

The Barcelona conference of 27–8 November 1995, when the fifteen EU members and twelve non-members (Morocco, Algeria, Tunisia, Egypt, Israel, the Palestinian Authority, Jordan, Lebanon, Syria, Turkey, Cyprus and Malta), discussed free trade and closer cooperation in the region, mentioned security only in its preventive, political and diplomatic aspects, assuming that the defence mechanisms would soon follow. 'They now have to be addressed openly, without the ambiguities and misunderstandings of the past.'[28]

In sum, the evolving mechanism of the EU's Common Foreign and Security Policy must synthesise and harmonise the foreign policies of EU member states so that they will not independently return to old and obsolete policies and practices seeking to create 'exclusive spheres of influence', 'antagonistic alliances' and 'special relationships'.

## NOTES AND REFERENCES

1. R. Aron, *Peace and War: A Theory of International Relations*, New York, Doubleday, 1966. Aron analyses international relations in the following sequence: theory, sociology, history and praxeology.
2. The terms are self-explanatory with the exception of 'praxeology' which for Aron entailed the analysis of contemporary events and the generation of policy recommendations.
3. P. J. Vatikiotis, 'The Mediterranean in Historical Perspective. Cultural, Ethnical and Religious Diversity' in Jelisaveta Stanojevich *et al.* (eds), *To Ellinikon. Studies in Honour of Speros Vryonis Jr,* Vol. 2, New York, Aristide D. Caratzas, 1993, p. 458.
4. Ibid. pp. 464–5.
5. For these and other useful statistics illustrating the North–South gap in the Mediterranean see Roberto Aliboni *et al.*, 'Cooperation and Stability in the Mediterranean: An Agenda for Partnership', *International Spectator,* vol. 29, no. 3, 1994, pp. 5–20.
6. Ibid.
7. For most comprehensive and analytical data on the comparative status of the human condition across the planet see *Human Development Report 1996*, United Nations Development Programme (UNDP), New York, Oxford University Press, 1996.
8. For this type of thinking see Samuel P. Huntington, 'The Clash of Civilizations?', *Foreign Affairs*, 72, Summer 1993 pp. 22–49.
9. See Eric A. Nordlinger, 'Soldiers in Mufti: The Impact of Military Rule Upon Economic and Social Change in the Non Western States', *American Political Science Review*, vol. LXIV, no. 4, 1970.
10. See for example E. H. Carr, *The Twenty Years' Crisis, 1919–1939: An Introduction to the Study of International Relations*, London, Macmillan, 1939; Hans J. Morgenthau, *Politics Among Nations: the Struggle for Power and Peace*, 5th edn, New York, Knopf; 1973; George F. Kennan, *Realities of American Foreign Policy* , Princeton, NJ, Princeton University Press, 1954; Henry Kissinger, *A World Restored: Metternich, Castlereagh and the Problems of Peace*, 1812–1822, New York, Grosset & Dunlap, 1964.
11. For a pioneering work of this school of thought see Robert O. Keohane and Joseph S. Nye, Jr, *Power and Interdependence: World Politics in Transition*, Boston, Little, Brown, 1977.
12. See Samuel P. Huntington's much discussed and highly controversial article 'The Clash of Civilizations?', op.cit.

13. Ibid. For the authors' highly critical response to the Huntington thesis, see Theodore A. Couloumbis and Thanos Veremis, 'In Search of New Barbarians: Samuel P. Huntington and the Clash of Civilizations', *Mediterranean Quarterly*, vol. 5, no. 1, 1994, pp. 36–44.
14. T. Veremis, 'International Relations in Southern Europe' in John Loughlin (ed.), *Southern European Studies Guide*, London, Bowker-Saur, 1993, pp. 207–210.
15. Ewan Anderson and Dominic Fenech, 'New dimensions in Mediterranean Security', in R. Gillespie (ed.), *Mediterranean Politics*, vol. I, London, Pinter, 1994, pp. 13–15.
16. Epaminondas Marias, 'Why Europe Must Now Look to the South', *European Brief*, vol. 2, no. 4, 1995, p. 22.
17. P. Henson: 'The European Union and the Mediterranean'. (Workshop Report, University of Reading, 17–18 June 1994), *Orient*, vol.4, 1994, pp. 531–9.
18. T. Couloumbis and T. Dokos (eds.), *Arms Control & Security in the Middle East and the CIS Republics*, Athens, ELIAMEP, 1995.
19. Geoffey Kemp, Forever Enemies? *American Policy and the Islamic Republic of Iran*, A Carnegie Endowment Book, 1994, pp. 36–7.
20. Shireen T. Hunter, *Turkey at the Crossroads: Islamic Past or European Future*, Brussels, CEPS Paper no. 63, 1995, pp. 103–4.
21. Ibid. See also Rosemary Forsythe, *The Politics of Oil in the Caucasus and Central Asia*, Adephi paper no. 300, London, IISS, 1996.
22. Alexis Alexandris, 'Tourkiki exoteriki politiki. Valkanike, Kaukasos kai Kentriki Asia', in T. Veremis (ed.), *I Tourkia Simera*, Athens, Papazissis-ELIAMEP, 1995, pp 470–3.
23. Nicolai A. Kovalsky (ed.), *Russia: The Mediterranean and Black Sea Region*, Moscow, Russian Academy of Sciences, Institute of Europe, 1996. See also on wider issues Geoffey Kemp, 'Security and Stability in the Persian Gulf: The Role of Iran and Iraq', Halki International Seminars: Arms Control Non-Proliferation, ELIAMEP, September 3–10, 1994.
24. Yannis Valinakis, 'Southern Europe Between Defence and New Threats: The View from Greece', in Roberto Aliboni (ed.), *Southern European Security in the 1990s*, London, Pinters, 1992, p. 60.
25. William H. Lewis, 'The Challenge of Failed States', Conference on Mediteranean Security in the Post-Cold War Era, AFSOUTH-Naples, 27 March 1995.
26. Richard Latter, *The Mediterranean Security Crisis*, Wilton Park Paper 103, July 1995, p. 14.

27. Michael Stuermer, 'Clausewitz and the Return of History at the End of the 20th Century', in A. Clesse *et al.* (eds), *The International System After the Collapse of the East–West Order*, Martinus Nijhoff, London, 1994, p. 263. Also, Roberto Aliboni, George Joffe, Tim Niblock (eds) *Security Challenges in the Mediterranean Region*, London, Frank Cass, 1996.
28. Guido Lenzi, 'European Security and the Mediterranean', in Alberto Bin (ed.), *Cooperation and Security in the Mediterranean. Prospects after Barcelona*, Malta, Mediterranean Academy of Diplomatic Studies, 1996, p. 33.

# 1 The European Union and Mediterranean Member States

Neville Waites and Stelios Stavridis

## INTRODUCTION

Any study of the foreign policies of Mediterranean EU Member states has to take account of the Union's impact on their purposes, methods and achievements or failures. The European Union's main impact on Mediterranean Member states has been to create elements of a common experience of economic and political life in the late twentieth-century world where otherwise the forces of friction predominate over those of harmony. The original European Community had only France and Italy as Mediterranean Member states. But the other Mediterranean states already felt the beginnings of a common experience from the early 1960s when they undertook to negotiate association or cooperation agreements with the EC (except for Albania, with its then closed Marxist-Leninist system, and Libya, with its oil-based economy and post-1969 revolutionary challenge to Western principles).[1]

From 1972 the creation of the Global Mediterranean Policy signified an EC belief in its power to manage a common economic experience in the area, and also an optimism regarding predictable economic development for all states in the Mediterranean and the wider world. This optimism was severely shaken by the oil crises of 1973 and 1979. But the sense of common experience was extended through the EC's Mediterranean enlargements to include Greece, Portugal and Spain in the 1980s, and has culminated in the Barcelona Conference's Euro-Mediterranean Partnership of November 1995 with all Mediterranean states, but still excluding Libya owing to international sanctions following accusations of terrorism. The common experience was felt more intensely by Member states, as a result of their transition from the EC to the more integrated European Union, than by non-Member Mediterranean states. But the latter's trading links with the European Union by the 1990s were much stronger than those with any other part of the world. Irrespective of cultural or political differences, they decided at Barcelona in 1995 to work for a free trade area with the EU between 2010 and 2015 as well as to cooperate in various socio-political undertakings of the new Euro-Mediterranean Partnership. However, the more developed process of convergence among EU Mediterranean Member states requires closer examination. In the pursuit of a European Union Common Foreign and Security Policy (CFSP) how far and under what conditions does the EU common experience of membership overcome friction generated by differing national interests?

## THE IMPACT OF COMMON EXPERIENCE

There are four distinctly positive features of common experience that influence foreign policies of EU states. Firstly, a fundamental condition of EU Membership is a democratic political system. When Winston Churchill commented that democracy was the worst system of government except for all known alternatives his words were appreciated as an endorsement of Western practices. But they could also serve as a salutary reminder of the strengths and weaknesses of various forms of democracy. None has achieved the status of a perfect model and today, more than in Churchill's time, much attention is given to debate about a 'democratic deficit' at the level of states as well as at the level of the European Union. Nevertheless, there is a consensus that democracy requires a government to be responsible to a parliament that is regularly elected in a form that allows for free debate and change in personnel and policies. Any democratic system therefore requires foreign policy, as well as other policies, to come under public scrutiny of its efficiency and to win popular approval of its direction. For example, British entry into the European Communities was debated and approved by Parliament in 1972, then reaffirmed by referendum in 1975.

For Members of an institution such as the European Union, as it became in 1993, domestic political support is therefore required for EU agreements. No Member state has yet lapsed from democratic practices, but it is assumed that such a lapse would be sanctioned by suspension or exclusion from the EU.

For some Members democracy was established long before entry, although the recent history of France, Germany and Italy gave real significance to their achievement of that condition. Its importance was enhanced even more by the enlargements of the 1980s, which were only possible through adoption of democratic systems of government in Greece, Portugal and Spain. That process involved opening governments to the judgement of mass domestic electorates. Foreign affairs ministries had to move away from exclusive traditional practices involving nepotism and privileged official appointments to develop a more efficient professionalism. Their work had to adjust to the interpenetration of domestic and foreign affairs which was greatly accelerated by involvement with Brussels, especially as the EC/EU was primarily a civilian power. All foreign ministries of new Members were restructured. The most striking example was when the Foreign Ministry in Athens opened an ultra-modern new wing in 1979, shortly before Greek entry into the EC, indicating adjustment to a new international role. For similar reasons, to facilitate coordination, different traditions of Foreign Ministry organisation among Member states have converged to a more standardised mixture of the functional approach, traditionally preferred by Italy as well as Germany, Belgium and the Netherlands, with the geographical approach, traditionally preferred by France, as well as Britain and Portugal.[2]

The second feature of the common experience was pressure towards convergence derived mainly from agreements to pursue common policy directions. Although for many years such common policies were confined to the economic sphere, the experience of collecting standardised data to enable the Brussels Commission to represent all Members' interests in trade negotiations with non-Members, ranging from the USA and Japan to African, Caribbean, Pacific and Mediterranean states, led to an awareness of EC/EU collective interests differing from those of states who were otherwise military allies in the Cold War period and beyond. Rivalry in world markets for steel, grain and meat might be classified as 'low politics', but it generated a team spirit that had spillover effects on 'high politics' as shown by efforts to pursue political cooperation which escalated to the Maastricht treaty agreement to create a European Union CFSP. Significantly, many of these efforts were concentrated on foreign and security policy problems in the Mediterranean area and, however limited their aims and achievements, they naturally had a great impact on EC/EU Mediterranean Member states.

The third feature of a common experience is the most easily quantified, namely the tremendous increase in intra-European Union trade. This increase has affected all Members. Some had only between 30 per cent and 40 per cent of trade with the EC/EU before becoming Members. Now it is the norm for Members to have between 60 per cent and 70 per cent of their trade within the EU. It can be argued that such visible trade statistics do not provide a complete picture when, for example, most of the very important invisible trade of one EU Member, the United Kingdom, is with non-Member states, notably those in the Pacific Rim, but this point is not applicable to the other Members who have a much lower level of invisible trade.

The essential point is that, while olive branches are not always evident in intra-EU negotiations, olives have become standard fare in shops and on dining tables in even the most northern regions of Europe. While Germans may not easily see the British point of view, they use British windscreen wipers on their VW cars to see more immediate dangers, and are the leading trading partners of all EU Members.

The fourth feature of a common experience derives from EC/EU institutional development during a twenty-year period. Much has been written about the power struggle between the Commission and the European Council which was established in 1975 and with the Council of Ministers prior to that (for example the 1965 crisis which led to the 1966 Luxembourg Compromise). But insufficient attention has been given to the important harmonising effects of each Member state, whether large or small, experiencing higher and wider levels of international power through the European Council's system of a rotating presidency supported by the troika of past, present and prospective presidents.

Insiders at Brussels are often sceptical regarding the efficiency of the Council and especially its troika system, but as a means of enhancing Members' practical participation and sense of belonging, at the official level at least, it is an invaluable source of strength for the European Union in all its branches and projects.

## FRICTION BETWEEN INTEREST GROUPS, REGIONS AND STATES

The European Union is well known for its difficult debates and negotiations leading to awkward compromise policy decisions. The friction was also evident within states, for example, in the French referendum on the Maastricht treaty in 1992 when most rural interest groups that were traditionally believed to benefit from EC/EU policies paradoxically voted 'No', but were narrowly outvoted by urban interest groups who voted 'Yes' in spite of being better placed to adapt to world market forces. Friction between states is often assumed to be predictable, such as that between Britain and France, but their links with Caribbean and African suppliers led them both to support EU policy on bananas. Similarly, British bitter fishing disputes with Spain did not prevent their cooperation on banana trade policy. Nevertheless, some economic friction can lead to serious violence, such as French obstruction of imports of Spanish wine or British lamb, or British obstruction of foreign fishermen.

Turning to high policy, there is often friction between the security interests of EU Members. The Schengen agreement on frontier cooperation has been difficult to implement for Italy and impossible for Britain and Ireland owing to disputes over controls of animals, criminals, drugs and immigrants arising from differing conceptions of internal security policy. Similarly, the role of Western European Union (WEU) as an EU security organisation is still disputed between those favouring European autonomy and those favouring total reliance on NATO. The widespread expectation that the end of the Cold War in the late 1980s would entail solutions to all major security problems has been replaced by widespread disappointment during the 1990s owing to the revival of parochialism and nationalism. Their revival generated new disputes and exacerbated old ones that were difficult to settle by international agreement. It was easier for the well established NATO system to act in the Gulf War and the Bosnian crisis than for WEU to implement the CFSP agreed in the Maastricht treaty. This was mainly because it was a new policy struggling to win consensus and support from EU Members who were influenced by a climate of national rivalry.

Concern over the evidence of friction within the EU led to a debate about weaknesses which were paradoxically increased by processes intended to strengthen it through further enlargement and integration. For example, it was argued that the EU had three fundamental weaknesses or 'deficits'.[3] Firstly, there was a deficit in democratic legitimacy. This took the form of concern regarding inadequate control and transparency of EU decision-making. But the main underlying concern was whether the EU was representing and defending the welfare and security interests of its citizens. It might be added that this was more fundamentally a deficit or decline in both Christian and Enlightenment conventions in Europe regarding individual rights and social cohesion, not merely an EU democratic deficit. Secondly, there was a deficit in institutional efficiency whereby increasing size and complexity had led to a deterioration in EU systems of decision making and implementation. Thirdly, there was a deficit in power distribution whereby in practice the resort to power politics prevailed over the EU theory of federal balance between municipalities, regions, states and Union. These deficits were there at the outset in 1957 but had been exacerbated to a critical degree by successive stages of EC/EU enlargement. Arguing that friction between various interests within the EU could not be reduced simply by promoting a common experience and seeking consensus, Joseph Janning is among those who have suggested that it would be realistic to pursue 'differentiated integration' and enlargement, accepting that there would have to be a core club of Members practising economic and monetary union, a Common Foreign and Security Policy, etc. Other Members would relate to some of these activities as they saw fit, but without being able to obstruct core club decisions.

EU Members' behaviour, for example during the process intended to lead to full monetary union by 1999, showed a general determination to stay within any 'core club' either by adjusting national policies or by seeking flexibility in conditions for club membership, or both. Northern Members' assumptions that Mediterranean states such as Italy and Spain would find Maastricht criteria for adoption of a single EU currency impossible to meet were resented and firmly rejected by the Italian and Spanish governments. Equally, it was proving very difficult for the long Inter-Governmental Conference (IGC) to devise new organisational systems that involved upgrading some Members at the expense of others in a search for greater efficiency. The attractions of a common experience and consensus politics appeared to have enduring qualities in spite of the passage of time and changing circumstances. Nevertheless, it was equally clear that the EC/EU rhetoric of integration had diverged dangerously far from reality by the 1990s. How did this problem affect Member states in the Mediterranean area?

## MEDITERRANEAN FORCES FOR BOTH UNITY AND DIVISION

The Mediterranean area had for many years generated its own kind of rhetoric regarding a common experience based on history and geography. Academically it was made respectable by Fernand Braudel. Politically it generated speeches and proposals that sought peace and cooperation in the area as a fundamental common purpose, even if they were directed against particular states or leaders who were identified as troublemakers. This rhetoric served various unifying projects ranging from 'Eurafrica' in the 1950s, thwarted by the process of divisive decolonisation, to the Euro-Mediterranean Partnership of the 1990s, threatened apparently by North–South cultural and economic cleavages amounting, according to Samuel Huntington, to a 'clash of civilisations'.[4] In the 1990s the prevalent popular perception in Europe of the Southern Mediterranean was that of an area of problems and threats.[5] It was an area attracting only about 3 per cent of the world's foreign direct investment.

Yet the grand theories of North–South conflict ignore or conceal the realities of North–North friction between, for example, Italy and France or Britain and Germany, equally the importance of South–South friction between, for example Egypt and Libya or Algeria and Morocco. Such theories also underestimate the extent of North–South sympathies that not only prevent conflict but also underlie the impressively consistent efforts of EC/EU states individually and collectively to develop harmonious relations across the Mediterranean. Why has there been such an EC/EU dynamic favouring its Mediterranean policy?

The 1957 treaty of Rome made provision for association and cooperation agreements which were focussed on Mediterranean Non-Member States (MNMS) in the 1960s and led to a more ambitious Global Mediterranean Policy by 1972. The early impetus came from French efforts to preserve relations with newly decolonised states in North Africa. They were supported by Italy, but also by Germany, thereby creating an EC consensus. The progress in 1972 was timed to accommodate British Mediterranean relations with states such as Cyprus, Egypt and Jordan within agreements on the first EC enlargement. The move accorded with a climate of optimism regarding world affairs and the power of the EC. Détente in the Cold War was moving towards the Helsinki process. The United Nations had expanded to welcome many newly decolonised states and their economies were expected to develop rapidly.

The prospects of economic development among MNMS were regarded as greater than those of African, Caribbean and Pacific (ACP) states. The Global Mediterranean Policy's trade and aid arrangements were correspondingly more modest and defensive, especially in view of the similarity of traditional agricultural products on both Northern and Southern shores of the Mediterranean. Traditional industries such as cotton, silk and leather were also a source of rivalry between Members and Associates. Thus there developed ambivalent attitudes among EC citizens towards MNMS. But policies of the EC and Member states always tipped the balance towards favourable relations.

This remained true even when the international climate of optimism was soon shaken by the Yom Kippur war and first oil crisis of 1973–74. Growing unemployment in Europe led to tighter immigration policies and even repatriation programmes, especially directed at Mediterranean foreign workers. Nevertheless, trade and financial protocols were concluded by 1976–77 with the Maghreb and Mashrek states. Moreover, the world multi-fibre agreement (MFA) made at that time to protect traditional textiles producers was designed to restrict Mediterranean states such as Greece, Turkey and Tunisia much less than those of the Pacific rim.

Meanwhile, sudden regime changes in Greece, Portugal and Spain attracted energetic support for democratisation from EC states that soon led to negotiations for the second and third EC enlargements. Whatever the clashes of economic interests between old and new Mediterranean Members, over products like wine, olive oil and fruits, there was never any doubt about the decision to enlarge the Community. In fact, this was so clear that the Greek threat to veto Portuguese and Spanish entry in 1986 should be seen as a procedural leverage to secure extra financial support through Integrated Mediterranean Programmes (IMPs) rather than as a policy of making the EC exclusive.

This process was also influenced by concern for Western security as East–West tension reached a new peak in the early 1980s. It meant that the issue of EC membership became embroiled in the question of NATO membership. For Greece it was a matter of possible withdrawal for PASOK in the 1980s in spite of Greek membership of NATO since the 1950s. For Spain especially EU Membership made NATO more palatable during the debate that took place prior to the 1986 referendum on continued membership as Spain had joined NATO only in 1982, although there had been close links with the Americans since the 1950s. This made EC–US relations highly sensitive and sometimes there was friction between security and economic interests that led to apparently low policy issues being treated as vital interests involving sovereignty, for example when the US tried to prevent EC companies from building a gas pipeline from the Soviet Union. Further security and economic problems were generated by the 1979 Iranian revolution. This created a second oil crisis causing massive inflation followed by a deep economic recession in the early 1980s.

Mediterranean states were by now deeply divided in at least four important ways. Firstly there were three who were new Members of the EC/EU, competing directly with MNMS over products such as citrus fruits and with older Member states over products such as wine and olive oil. Secondly there were sharp divisions between oil producers like Algeria or Libya and oil consumers like Morocco or Jordan or the EC Member states. Thirdly there were differing positions in the East–West confrontation in so far as Algeria, Libya and Syria obtained armaments from the Soviet Union and Greece and Spain were sometimes sceptical towards American hectoring and sabre-rattling whereas Italy and France tended to follow the Washington line. Fourthly the spillover effects of the Iranian revolution sharpened cultural divisions between states with Christian traditions committed to liberal democratic modernity and states with Islamic traditions committed to retaining tribal, family and communal values. The secular political systems of Islamic societies were challenged from within by militant Islamists drawing inspiration from Iran to reject secularism which they portrayed as a form of Western corruption. European societies by the 1980s contained over 4 million immigrants from Mediterranean Islamic societies and that led to fears of their subversion or at least disruption through terrorism. It was in this fractious climate that trade and aid agreements between the EC/EU and Maghreb and Mashrek states had to be renegotiated in a period of economic recession. The EC/EU Mediterranean policy was patched up with a little extra funding and a shift from tariffs to non-tariff barriers such as calendar restrictions and quotas, cultivating self-discipline among MNMS and applying band-aid dressings to their wounds from the trade war by providing them with some additional finance.

A new phase of EC/EU Mediterranean policy followed the end of the Cold War and the far-reaching security, political and economic changes which were especially important in opening up Eastern Europe to the EU and NATO. Since 1989 there has developed a symbiotic relationship between EU policies towards the Mediterranean and Eastern Europe in which the latter has gained a prime, but not exclusive importance. The symbiosis can be illustrated by the grants to MNMS of approximately 5 billion ECU agreed in 1995, following intense pressure from Mediterranean EU Members, compared with approximately 7 billion ECU already agreed for Eastern European non-Member states. Germany, contributing over one-third of the EU budget, gave higher priority to Eastern Europe whilst recognising the importance of a viable EU Mediterranean policy, and this view reflected an EU consensus. Another illustration of the symbiosis was the way that the advantage of the Mediterranean states, Cyprus and Malta, in gaining an EU commitment in March 1995 to negotiate on their application for Membership within six months after the end of the Inter-Governmental Conference to deal with restructuring, was counterbalanced a few months later by a similar commitment to applicants from Eastern Europe.

Progress was also made by other MNMS through an EU Mediterranean initiative launched under the Greek presidency at the Corfu European Council meeting in June 1994, significantly developed under the subsequent German presidency and defined at the Essen Council meeting in December as 'a programme to establish a Euro-Mediterranean Partnership to promote peace, stability, prosperity and cooperation in the region'.[6] The Cannes summit under the French presidency in June 1995 created the framework for the Barcelona Conference in November to bring together all EU and MNMS (except Libya) to launch the Euro-Mediterranean Partnership.

The first originality of the Barcelona Conference was that it brought together all the foreign ministers of the 15 EU states and the 12 MNMS. It complemented existing trade and aid arrangements managed from Brussels, which took the form of new bilateral Association Agreements with Tunisia and Israel in 1995, with others to follow for the other MNMS and the Palestinian Authority. The grant aid total of 4,685 million ECU for a five-year period amounted to double that of the previous agreements. A further development was the aim to create a free trade area between 2010 and 2015. A second original feature of the Conference was the adoption of three aspects ('volets') of action to look into political, economic and social issues. A third innovation was to create a Euro-Mediterranean Committee for the Barcelona Process consisting of EU Council troika officials and those representing the twelve

MNMS. The Committee would implement and monitor the agreed follow-up meetings and work on each 'volet' at both ministerial and official levels. This was a substantial advance on earlier experiments in Mediterranean cooperation which had little follow-up and depended for progress on constant ministerial action. This feature impressed some observers who had been disappointed with the relatively cautious financial and economic agreements at Barcelona. As the three 'volets' produced a flow of meetings in 1996 on industry, energy, information technology, drugs, crime and culture, it was argued that the political spillover effects of Barcelona, by creating public and private sector fora and a culture of multilateral discussion and cooperation, might become the most important aspect of the Euro-Mediterranean Partnership.[7]

This progress in Mediterranean cooperation needs some explanation. A view frequently encountered is that such ambitions were natural for reasons of geographical proximity and shared history.[8] It is also clear that the internal political and administrative dynamics of the EU favoured initiatives for the Mediterranean area in accordance with its symbiotic relationship with Eastern Europe in EU policies. Another view is based on economic interests, especially in supplies of energy and raw materials produced or transported in the Mediterranean area. But such an approach has to take account of the dramatic decline in relative importance to the EU of MNMS trade. There are many alternative sources of energy and raw materials around the world. The development of the EU single market and the diversification of EU external trade brought, for example, a decline in French imports from OPEC states from 17 per cent of all French imports in 1976 to 3.6 per cent by 1992. By that year French trade with the whole of the Middle East and North Africa was exceeded by French trade with Spain alone. Moreover, trade with France of the five Maghreb states by 1995 was matched by that of the three new Members of the EU, Austria, Finland and Sweden, even before their entry.[9] A cynical conclusion to draw from these trends is that limited economic interest accounted for the limited funding for EU Mediterranean policies even under the Barcelona Process. Apart from restricted collective EU finance, it has been pointed out that individual Mediterranean Member states had not put their money where their mouths were. Spain's promise in 1986 to raise foreign aid to 0.36 per cent of GDP contrasted with the actual achievement of only 0.20 per cent by 1994.[10]

All that evidence, however, should be seen in the context of a dominantly liberal economic value system in the world since the 1970s according to which public investment unrelated to market trends might be considered futile and trade was given vastly more importance than aid. By the 1990s most value was attached to indicators of direct private foreign investment. Unfortunately, in 1992 for example, the whole of the Middle East and North Africa only received 2.1 per cent of world direct private foreign investment.[11] The picture was apparently one of MNMS trade imbalance, heavy debts, low investment, satellite-type relations with Europe as suppliers of energy, raw materials, labour and cheap industrial components. Nevertheless, sharp falls in their birth rates towards 2 per cent per annum confounded analyses of only a decade earlier. And the trend towards extensive global production that transformed economies not only in the Pacific rim but also in Ireland and Portugal had great potential benefits for economies in the Mediterranean area. Multinational companies were stimulated by the lowering of tariffs and non-tariff barriers after GATT agreements such as the Uruguay Round, and by the lowering of transport and communication costs through information technology. More sophisticated differentiation in production and marketing on a world scale was unlikely to leave populous Mediterranean societies on the margins of the global economy for long. The EU plays a major part in global economic competition with over 40 per cent of world trade. But it has seen its influence diminish in some areas. For example, its ACP policy has been weakened by North American Free Trade Area influence in the Caribbean and by that of Pacific rim economies in their region. These developments have stimulated the EU to defend its influence in the Mediterranean area where it has natural advantages. But a purely economic agenda for the Mediterranean had shown its inadequacies by the 1990s.

The nature of the 1995 initiative at the Barcelona Conference, with its three 'volets', could also be explained with reference to EU security concerns in the Mediterranean area. They were not primarily of a classical military kind after the end of the Cold War, although nuclear, chemical and conventional weaponry still had their place in various scenarios of potential and actual conflict, as shown by the Gulf War and the Bosnian crisis. Uncertainties about security after the Cold War led to the decision at Maastricht in 1991 to develop an EU CFSP with Western European Union (WEU) as its military organisation. But the resilience of NATO in general and American power in particular, as revealed by the Gulf War and subsequent crises such as in Bosnia, contrasted sharply with a lack of unity and sense of direction among EU Member states. They therefore resorted largely to solving security problems through NATO in the short term. Mediterranean Member states maintained their support for WEU and participated in the development of the Eurofor and Euromarfor land and sea-based military organisations. But Northern EU states were more sceptical towards WEU, either through traditions of neutrality, such as Ireland and Sweden, or through traditions of Atlanticism, such as Britain and the Netherlands. The development of German policy towards WEU was likely to be both a sensitive and a crucial influence, but it remained ambivalent between WEU and NATO.

In the Mediterranean area the EU had urgent security problems such as the control of drug-trafficking, large-scale migration and terrorism. In view of the very nature of these problems, which were not easy to combat with armies and navies, the EU states sought wide-ranging solutions involving cooperation with the MNMS from which some of these problems emanated. A purely security agenda had been proposed by NATO in February 1995 to combat Islamic militias and terrorists but the Arab states found this approach inadequate.[12] The Euro-Mediterranean Partnership established in November 1995 with three 'volets' had the advantage of using EU economic and financial assets and also taking account of its ambition to be more than a civilian power. Many European leaders shared the assumption that promotion of economic growth was the prime means of reducing social and international friction of the kind witnessed in the Mediterranean area in the 1990s. But they were also aware of the cultural and religious challenge to Western economic values especially from within Islamic societies. They therefore recognised the importance of the third 'volet' to promote cultural understanding and cooperation.

There was no guarantee that its more ambitious structure would make the Barcelona Process for Euro-Mediterranean Partnership more successful than such ephemeral experiments as the Five plus Five cooperation on security issues or the Mediterranean Forum for wider ranging talks or the Conference on Security and Cooperation in the Mediterranean, which was a weak offshoot of the CSCE that had produced the East-West agreements through the Helsinki process. It has been pointed out that some of the aims and characteristics of the Euro-Mediterranean Partnership were perhaps influenced by the Helsinki process, not least the device of working through several interrelated 'volets' or 'corbeilles' (baskets).[13] The Barcelona aims were more complex and geared more to low politics. The achievement of gathering ministers from almost all the twelve MNMS was considerable. Equally the energetic participation of the Northern EU states, including Britain and Germany, in a Conference that left out the United States was a measure of the real importance attached to the Mediterranean area by all EU Members. Its importance lay in awareness firstly of long-term prospects in an area with important assets for the EU in global economic competition, and secondly of the integral relationship between political order and social progress both North and South of the Mediterranean, which involved making compatible the religious cultures and national aspirations of the whole area.

## CONCLUSIONS

The European Union already had a long enough history and sufficient size and power by the 1990s to have formed an important part of the common experience of all states in the Mediterranean area. Its influence had been both unifying and divisive. There were many intractable problems to be overcome. Firstly, several states were in exceptional situations. Libya was excluded from normal communication by international sanctions. Algeria was severely handicapped by a long civil war, and in a different sense the Israeli–Palestinian relations still resembled those of civil war, albeit in a 'cold' form since the 1993 Oslo agreements had established the Palestinian Authority. Cyprus was still divided in spite of successful economic and political reorganisation in the south of the island since 1974 and UN efforts supported by the EU to promote reunification with the northern sector.

Secondly, many MNMS were still struggling to find acceptable means for mass public participation in political power, even if not democratisation on the lines of EU states. Moreover, assertions and infringements of human rights in states ranging from Morocco to Turkey frequently influenced EU Members' foreign policies. This was apparent especially since the July 1987 entry into force of the Single European Act which enabled the European Parliament to use powers of 'assent' to insist on human rights conditions for progress even with purely economic EU agreements with MNMS. There was a need in the whole Mediterranean area to reduce political violence and also social friction arising from unemployment and large scale migration.

Thirdly, considerable economic reform was still needed to create wealth and raise living standards in southern regions of the EU and in most of the MNMS. There was still uncertainty about how to reconcile EU purposes, involving intervention and control, with liberal conventions in the global economy. Growth and efficiency in terms of economic production at world market prices often meant a fall in living standards for many individuals and groups. This problem was part of the larger one of unbalanced economic interdependence in the Mediterranean area.

Speeches by political leaders of Mediterranean states are notable for their convention of beginning with references to the common experience of a glorious history and attractive geography, and ending with reference to common aspirations for cooperation, peace and prosperity. But the devil is in the detail of such speeches, where points of friction and accusation become apparent. Nearly all Mediterranean states have had periods of illustrious history, but often at the expense of neighbours. Equally, all have had periods of violent subjection to enemies in the twentieth century. That gives them a sense of great pride but at the same time a sense of great vulnerability. They understand each other very well but trust each other very little.[14] Perhaps the most encouraging feature of EU Mediterranean policy is that it has been a collective creation rather than depending on leadership from ephemeral governments or individuals. That indicates its strength and durability as a permanent feature in the foreign policies of all EU states and others in the Mediterranean area.

## NOTES AND REFERENCES

1. Accounts of the detailed development of EC/EU Mediterranean policy can be found in R. Pomfret, *The Mediterranean Policy of the European Community*, London, Macmillan, 1986; Eberhard Rhein, 'Europe and the Mediterranean: A Newly Emerging Geographical Area?' *European Foreign Affairs Review*, vol. 1, no. 1, July 1996, pp. 79–86 and Geoffrey Edwards and Eric Philippart, 'Mare Nostrum: the European Union and the Mediterranean in the 1990s and Beyond', paper given at 5th biennial international Conference of the European Community Studies Association at Seattle, 29 May – 1 June 1997 produced at the Centre of International Studies, Cambridge, UK.
2. Zara Steiner (ed.), *The Times Survey of Foreign Ministries of the World*, London, Times Books, 1982, pp. 25–31.
3. Josef Janning, 'The Emerging Political System of Europe', paper given at ELIAMEP international Seminar, Halki, Sept. 1996, published in Spanish as 'EI Sistema politico emergente en Europa', *Meridiano*, CERI (Madrid), no. 2, 1995, pp. 4–9.
4. Samuel P. Huntington, 'The Clash of Civilisations?' *Foreign Affairs*, 72, Summer 1993 pp. 22–49.
5. Janning, op. cit.
6. European Council, Essen Summit 9–10 Dec. 1994, Presidency Conclusions, p. 11, EU SN 300/94.
7. George Joffe, 'Partnership with the EU: What's in it for the Maghreb States?' Chatham House Middle East Programme, London, RIIA, March 1997, pp. 1–4.
8. P-M. de La Gorce, 'La France et le Maghreb', *Politique étrangère*, Winter 1995–6, p. 27.

9. Neville Waites, 'France and the Post-Colonial Mediterranean Region', University of Reading *Discussion Papers in European and International Social Science Research*, no 53, Sept. 1994, pp. 8–9; statistics are drawn from *Direction of Trade Yearbooks*, Washington, IMF.
10. Alfred Tovias, 'The EU's Mediterranean Policies under Pressure', in R. Gillespie (ed.), *Mediterranean Politics*, London, Pinter, vol. 2, 1996 p. 9.
11. Joffe, op.cit. p. 3.
12. B. Kodmani-Darwish, 'La France et le Moyen-Orient: entre nostalgie et réalisme', *Politique étrangère*, Winter 1995–6, pp. 994–6.
13. Joffe, op. cit. p. 3.
14. For a discussion of the implications of this point see Nicholas Rengger, 'The Ethics of Trust in World Politics', *International Affairs*, vol. 73, no. 3, July 1997, pp. 469–87.

# 2 France's Foreign Policy in the Mediterranean

Laurent Meyrede

La Méditerranée, c'est un continent où 'les bateaux naviguent; les vagues répètent leur chanson; les vignerons descendent des collines des Cinque Terre, sur la Riviera génoise; les olives sont gaulées en Provence et en Grèce; les pécheurs tirent leurs filets sur la lagune immobile de Venise ou dans les canaux de Djerba; des charpentiers construisent des barques pareilles aujourd'hui à celles d'hier... Et cette fois encore, à les regarder, nous sommes hors du temps'.

Fernand Braudel

## INTRODUCTION

After the return to power of the Right in 1993, the Mediterranean appeared to be a priority of French foreign policy. The commitment of the Balladur government (1993–5) to obtain a substantial increase in European Union assistance to the Mediterranean, and the choice of Morocco for the first foreign visit of President Chirac are but two important indications of this renewed interest for the *mare nostrum*. Both events were widely noted in the media.

What are the aims of French foreign policy in this area? The success of these two initiatives could not characterise alone the real or declared aims of France's foreign policy in this region which is a crossroads of civilisations and of ambitious national ventures, including hegemonic cultural, religious and economic challenges.

The success of a foreign policy depends on having effective methods and means of expression and on support from a national consensus. This chapter, therefore, attempts to evaluate these elements in order to define better the prospects of success for French initiatives in the Mediterranean and the possibilities of synergy with other European partners.

## THE DEVELOPMENT OF FRENCH MEDITERRANEAN POLICY: THE ROLE OF THOSE IN POWER

In October 1995, both the announcement of a meeting (later cancelled) between President Chirac and Algerian General Liamine Zeroual on the occasion of the UN's 50th anniversary in New York, and the unjustified criticisms made during the summer of 1995 of the Juppé government in its handling of the threat emanating from terrorists linked to the Islamic Armed Groups (GIA), created a controversial debate within the political elite and public opinion. Both examples highlighted the plurality of Mediterranean policy approaches and their integration in wider French foreign policy. This plurality of approaches reflects the diversity of personalities influencing the elaboration of French foreign policy on the Mediterranean.

### A new generation of leaders: the generation of the Algerian War

With the departure of Mitterrand in May 1995, France turned a page of its contemporary history. Those serving the state who lived through World War II have been replaced by a new generation: that of the conscripts of the Algerian War and of those repatriated from North Africa during the period 1954–62. From the beginning of the 1980s, those in high public office, and notably in the army, were people who had served as young officers in Algeria or had experienced in France or North Africa the heartbreaks of the decolonisation process and, for some of them, even the passions of a commitment incompatible with the ultimate decision of General de Gaulle to leave Algeria.

In 1995, this transition was carried further by the accession to leadership of the French state of two men deeply attached to the Mediterranean in general and to its southern shores in particular. The first is President Chirac, who had chosen to go to French Algeria while he was doing his national service as a junior lieutenant. At that time Algeria was suffering from fratricidal violence during the war of independence. The second person is Philippe Séguin who, as President of the National Assembly and occupying one of the highest places in the order of French protocol, has substantial political influence. It should be noted that he was born and brought up in Tunisia.

*The influence of the Arab policy of General de Gaulle*

Current personalities on the political stage began to emerge in the 1960s. That decade was marked by General de Gaulle's ambition to give France a world influence. The Arab policy of General de Gaulle was one of the keystones of this ambition to be recognised as a great world power dealing on an equal footing with the United States and the Soviet Union.[1] France's interest in Arab countries focused essentially on the Mediterranean basin.

The two axes of French penetration in the Arab world were the Maghreb and the Lebanon. This policy of grandeur that the French would describe as 'l'exception française' is the biggest point of reference for leaders claiming a Gaullist heritage. The Gaullist inheritance represents, on the one hand, a significant political legitimacy in the French political landscape and, on the other hand, a fund of enduring sympathy with Arab leaders. This sympathy is still real to judge by some particularly cordial messages addressed by several of these leaders to Chirac after his accession to the Presidency.[2] Indeed, Arab newspapers gave a relatively favourable analysis of the implications of the arrival at the Elysée of a man who is an avowed 'Gaullist'.

The Mediterranean, and Arab countries on its shores, appear to have a special importance for French diplomacy in its strategy to remain a great power on the world stage. This appears to be the last remaining area of the world where France can have undisputed leadership. Indeed, the Mediterranean area still has special advantages for French foreign policy at a time when France finds it increasingly difficult to justify both its permanent member status on the UN Security Council and its nuclear tests; when reunified Germany has become an actor, perhaps a competitor, on the international diplomatic stage and a leader in the European integration process; when Spain and Italy assert themselves each year more and more as powers seeking more regional influence; and when successful American initiatives in former Yugoslavia and in the Middle East have somewhat eclipsed any moves from Paris.

*Personal relations*
Personal relations maintained by French leaders with Mediterranean countries' leaders have a little known but real influence on the formulation of French policy towards the *mare nostrum*.

The Cherifian sovereign Hassan II, who knows well the sensitivities of French society, has kept firm relations with French leaders, especially Gaullists, over several decades. These relations are sufficiently strong for the two countries to promote mutual cooperation and to lend assistance if necessary. French authorities have thus given valuable advice and support to the Moroccan armies campaigning in the Western Sahara in the 1970s and 1980s and have prevented, on several occasions, Hassan II's opponents on French soil from becoming a real threat to his power. More recently, pressures exercised on the journalist Gilles Perrault, the author of a very critical work on King Hassan II, shed light on the close bond between Rabat and Paris.[3] Morocco, for its part, has intervened militarily in several countries under French influence in Africa, notably at Shaba in Zaire in 1977–8 to defend the interests of Paris when they were seriously threatened and when a French military intervention would have been badly received diplomatically. One finds Moroccan military counsellors or technical engineers alongside former French officers in several presidential guards in countries of fundamental interest to Paris, for example in that of President Omar Bongo of Gabon. The friendship maintained by Hassan II and Chirac dates from the beginning of President Giscard d'Estaing's presidential term when Chirac first became Prime Minister.[4] This friendship was maintained during the subsequent Socialist presidencies.[5] It should be noted moreover that Chirac – after his first departure from the Hotel Matignon in 1976[6] – established an embryonic diplomatic cell within the 'Mairie de Paris'. With his position as Mayor, Chirac was able to maintain close contacts with foreign leaders for almost twenty years and was able to constitute a team of efficient specialists in African and Arab affairs. In this team, perceived by socialist leaders as a shadow diplomatic apparatus serving the Rassemblement Pour la République (RPR),[7] one found men of high quality such as Roussin,[8] an African affairs specialist, and more substantial groups such as those of Pasqua and Marchiani, leading Arab affairs specialists. Thus, even when he was far from the presidency or the government, Chirac was able to maintain a high-level knowledge of political affairs in the Mediterranean and particularly in North Africa. These relationships were reaffirmed in 1995 when President Chirac chose Morocco for his first official foreign visit. This highly symbolic choice indicated the importance of the Mediterranean in the orientation of French foreign policy. In the course of difficult negotiations for an association agreement with the European Union, Morocco was able to rely, on the one hand, on the consistent, discreet and effective support of French authorities and, on the other hand, on lobbying groups in Brussels who were aware of Paris–Rabat relations.

The Lebanese Prime Minister Hariri, who lived for a long time in Paris, where his children were educated, is also a personal friend of Chirac. The latter, when Mayor of Paris, had moreover undertaken an official visit to give his endorsement to Hariri when he became Prime Minister of the Lebanon.

The Tunisian President, Zine el Abidine Ben Ali, a former student of the French military school of Saint-Cyr maintains personal relations with several French leaders.

General Liamine Zéroual, elected president of the Algerian Republic in November 1995, does not know Chirac personally, but he maintains close relations with senators and deputies such as Yves Bonnet, former chief of the French counter-espionage service (DST).

The experience, ideological points of reference, relationships and personal sensitivities of the French leadership have a powerful influence on the orientations of French foreign policy, in the Mediterranean area. The Mediterranean for the French is perceived essentially as its western part. This is a result of human and historical bonds between the former metropolis and its colonies in the Maghreb. Underlying official actions and declarations there is a plurality of conceptions of what French foreign policy in the Mediterranean area should be. The Mediterranean policy of France is a compromise between these multiple conceptions of interests and threats.

Are the conceptions of state technocrats and public opinion as clear and coherent as those of French leaders? An analysis of the positions of the different ministries and the different currents of public opinion will highlight axes of convergence and divergence on French Mediterranean policy.

## THE DIFFERENT CONCEPTIONS OF FRENCH FOREIGN POLICY IN THE MEDITERRANEAN AREA

### French official conceptions: an interministerial approach

For French diplomacy, the principal stake in the Mediterranean area lies less in the prevention of inter-state conflicts than in the internal stability of Arab regimes, especially in the western Mediterranean. From a French viewpoint, the stability of these regimes determines, on the one hand, the security of the common space which is the Mediterranean and, on the other hand, the absence of disturbance within the important community of Maghrebi origin residing on French territory. The security risk relating to the Maghrebi community is included in the analysis of the Quai d'Orsay.[9] Indeed, even if for issues of international war and peace the western Mediterranean is less important than the eastern Mediterranean, it is clear that the Algerian situation could modify that priority rapidly and radically. The perception of a stabilisation, even a return to normality, of the Algerian situation after the 16 November 1995 presidential elections, was not confirmed by events in 1996. Violence still continued in Algeria; bomb attacks causing the death of many people did not cease. The Algerian press, controlled by the military authority's censorship, could not reveal the real extent of violence.[10] This contributed to falsifying the perception of some French academic experts. By their articles and participation in conferences they have influenced French government circles and also those of several other European governments. These expert analyses was used as a reference point to understand the complex Algerian situation. However, that situation was changing to the extent that non-Mediterranean European states observing the superficiality, even the absence of coherence in the work of some French academics, gave less credit to such analyses. As far as credibility was concerned, French diplomats came to regret that analyses of the situation in

Algeria made by these experts, who provided direction of thought in the field of French political science and international relations, revealed a cumulative process that resulted in such superficiality.

The diplomat of the Quai d'Orsay, in his attention to strategic risks in the Mediterranean area, is confronted with the problem of supporting Maghreb regimes in moves towards a measure of economic and political openness. These regimes are evolving in a context where, in the face of an Islamist challenge (which is sometimes violent), a rapid political transfer of power is not possible. It would be even harmful, according to consultants working specifically for the Quai d'Orsay. Arab leaders are compelled to make ideological concessions to the Islamists in so far as they do not change prerogatives and privileges of the clans who run the state. The tactics of the Maghreb leaders have the clear major objective of maintaining their power. For this purpose, Arab leaders develop a discourse, essentially for domestic consumption, which is against western values. The French authorities are aware of the necessity, even the value, of this discourse of variable geometry. This takes account of the limited room for manoeuvre of friendly Arab leaders in southern Mediterranean countries. This analysis would explain the conciliatory, or at least the discreet position of French diplomats witnessing actions against democracy and principles of human rights according to Western liberal standards.

This French attitude, if it is appreciated by Arab leaders sensitive to the principle of non-interference, is often seen by southern populations as complicity motivated by the need to obtain active cooperation in stopping migratory flows and in promoting business interests. These populations, compelled by the French government and other European authorities to accept a very strict procedure of entry visas, see the former colonial metropolis, on the one hand, as a hostile arrogant power and yet, on the other hand, as an attractive model of material success. This feeling of both attraction and repulsion is widespread among Maghrebi youth in general and, more particularly, among those seeking emigration.

Diplomats of the Quai d'Orsay who are in charge of foreign policy in the Mediterranean area have to be extremely careful to convince southern populations that their strategy is not an unconditional endorsement of political regimes which are losing credibility. French diplomats have to use an appropriate language to show that French foreign policy is not a struggle against Islamic politics, and certainly not against the Muslim religion. Such a confrontational perspective would have disastrous effects on French interests and the security of French nationals in Islamic countries, who would then become targets for public hatred under the influence of some extremist leaders.

Under the direction of M. Charles Pasqua between 1993 and 1995, the Ministry of the Interior was able to become a major actor elaborating its own French foreign policy in the Mediterranean area. The organisation of coordination meetings against terrorism between Maghrebi Ministers of the Interior and personal emissaries dispatched to Arab leaders highlighted Pasqua's interest in the Mediterranean space. This area has often been perceived, as have West African French-speaking states, as a natural 'chasse gardée' of France. The interest of Pasqua and other Interior Ministry leaders was linked to their geopolitical perceptions of the new risks coming from individual actors or non-governmental groups operating in the southern Mediterranean area: a growing potential of illegal immigration, the increase of drug traffic, notably from Morocco, and the extension of terrorist networks linked to Algerian Islamic armed groups – even their connection with delinquents in the suburbs of large French cities.[11] In the view of these highly placed officials, legitimately worried about national security, the foreign policy of France in the Mediterranean should be principally to obtain from southern regimes an active and unfailing collaboration in the fight against these new threats. Such cooperation could thereby legitimise compromises on human rights and democratisation.

The Ministry of Economy and Finance (known as Bercy) and more particularly the Directorate of External Economic Relations (DREE) consider the Mediterranean as an area where French firms achieve overall good results. In the five countries of the Maghreb, France is traditionally one of the three leading suppliers.[12] With the exception of Libya (where Italy and Germany together represented 42 per cent of the market in 1994), France achieves comfortable financial surpluses. The Mediterranean area represents also one of the most important financial risks for COFACE – the national organisation offering insurance to cover risks linked to exports – and the French Treasury.

According to economists, there can be no development, which is a necessary condition for success in the employment battle and to reduce youth frustration, without internal security and prospects of credit-worthy markets. These two elements are considered as absolutely necessary for the mobilisation of domestic savings and for attracting foreign direct investment. But what is the actual state of the economy in the Maghreb which is so important for French economic and financial leaders? Violence is ubiquitous in Algeria, the prospects for the succession to King Hassan II of Morocco are bad, and Tunisian society gives the impression of being stifled. To this generally depressing political picture one must add a flight of capital which reveals an absence of confidence among local investors.

The preference of Ministry of Economy and Finance experts, and of French businessmen, for endorsing southern Mediterranean regimes which are in office is based on a simple diagnosis: the devastating effects of terrorism, the absence of a clear, stable and open judicial framework, the lack of predictable long-term perspectives, and the lack of transparency in business life do not promote investment. This situation consequently prevents the durable development which is essential for a local economic take-off. To remedy the 'mal maghrébin', French businessmen and Bercy officials recommend the adoption of a progressive development programme based on commercial opportunity and a market economy. To reach this target, southern countries have to replace a centralised and bureaucratic system with a system which welcomes citizens as full actors in social and economic life. At the same time it would be an error to consider that French businessmen would reject strong government which they see as the best able to ensure debt servicing and to maintain internal security.

From the point of view of the Ministry of Economy and Finance, which participates in fixing the amount of the bilateral assistance to southern Mediterranean countries, French diplomatic action has to promote, first of all, internal stability, then the opening of markets and finally, structural adaptation to an environment with strong international competition.

## French public opinion: multiple interpretations

If French public opinion had traditionally a certain generosity and tolerance with regard to its neighbours on the southern shores of the Mediterranean, a change was clearly perceptible during the 1980s. Economic difficulties made French people very critical, even opposed to Maghrebi immigration. Electoral results obtained by the Front National, the extreme right party of Le Pen, revealed the extent of this fear of immigration.

Public opinion gradually developed a siege mentality which pressed the government to restrict or even close the channels of immigration. Political leaders on both the Right and the Left took account of this widespread popular demand. This applied to leaders of the Right in the government of cohabitation directed by Chirac under Mitterand's presidency (1986–8) and of the Left with the Cresson, Rocard and Beregovoy governments (1988–93). The introduction of entry visas for southern Mediterranean nationals arose from the concern of French political leaders to give pledges of firmness. From 1992 the prospect of the advent to power of Islamists in Algeria, and the increase in daily violence, led French public opinion to fear an invasion of Algerian 'boat people'. Leaders of opinion demanded that national authorities should close borders and also sustain the military regime in Algiers to prevent an Iranian scenario that would drive thousands of Algerians on the roads to exile. The French were convinced that their country would undergo the strongest pressure from Algerians fleeing Islamic chaos. Indeed most French people believed that the prime destination for Algerians seeking political asylum would be France.

In the early summer of 1995, even before bomb attacks conducted by Islamists in France took place, an opinion poll showed that the greatest fear of the French was the Algerian crisis and its consequences on French territory. As noted above, one of these worrying consequences would be mass immigration from Algeria. This fear of a massive immigration should be seen in the context of social and economic crisis which was particularly severe in the suburbs of large French cities. The terrorist attacks in the summer of 1995, attributed to the Islamists, shook Paris and strengthened a growing fear in French public opinion.

This fear soon gave rise to a wish to see the state show more firmness with regard to terrorists and to impose exemplary sanctions against individuals and foreign states supporting them. Several public opinion polls showed that for the majority of French citizens France's foreign policy in the Mediterranean area had to use efficient and concrete measures:

- to prevent any destabilisation in the southern Mediterranean that would be caused by the advent to power of Islamists hostile to the universal values defended and promoted by France;
- to protect France against 'the southern threat'.

This perception of the threat covered essentially three areas: Islamic terrorism on French territory or against French interests, the assassination of French citizens, migratory pressure and the subversion of populations of Maghrebi origins living in the suburbs of the larger cities in France.

This will to keep the French community of Maghrebi origins free from foreign influence is accountable, on the one hand, to its numerical importance and, on the other hand, to the poor understanding and the lack of dialogue during the last three decades between the traditional and new components of the French community. France provides a home to the largest community of Maghrebi origin in the European Union. Islam is the main common characteristic of this community. Otherwise it is heterogeneous in its opinions, its practices, its projects for the future, its relations with the Maghreb area and, finally, in its conceptions of French foreign policy in the Mediterranean.

During the Gulf War, French people of Maghrebi origin and North Africans living in France were extremely discreet and respectful of the positions taken by France while, at the same time, southern Mediterranean populations demonstrated with passion in the streets their endorsement of Saddam Hussein. The 'beurs' generation (women and men born in France of parents who come from North Africa) is not sufficiently organised politically to have a lobbying capability.

If this capacity were to be developed, the orientation of French foreign policy in the Mediterranean area would be unlikely to become a priority for such a lobby.

Even if this orientation were to become a priority, it is not evident that young "beurs" would feel particularly attached to their parents' or grandparents' homelands. Indeed, for this young generation, the main preoccupation is social and professional participation in a society with high unemployment. Nevertheless, this community, with a developing awareness of its duties and rights, is particularly sensitive to respect for human rights. These young women and men consider that French diplomacy, that of their own country, can no longer support unconditionally discredited regimes which blithely disregard the most elementary individual rights. The beurs wish to see in the foreign policy of their country towards the southern Mediterranean the following two characteristics: a firmness of language and a refusal to compromise with non-democratically elected regimes.

Advocates of radical Islam who operate in France within the Maghrebi community have developed a discourse that stigmatises the lack of legitimacy of the southern Mediterranean regimes. This discourse can find a favourable echo among unemployed youths searching for an ideal or commitment. Moreover, if in the eyes of these youths French foreign policy appears to accept or even support these southern Mediterranean regimes, it is probable that France will be made the object of rejection. France will then be considered a directly legitimate target for Islamic fighters. The actions of the young Khaled Kelkal, one of alleged perpetrators of the attacks in the summer of 1995, was particularly revealing of the fragility of the moral commitment of this young beur to the French community within which he did not feel sufficiently recognised and valued.

More generally, one can consider that for the large majority of the Muslim community in France, French policy in the Mediterranean has to avoid being seen as a struggle of Christian civilisation against political Islam and also an unconditional endorsement of corrupt regimes which lack real democratic legitimacy. French policy, in their view, should rather favour a process of political openness which is respectful of Arabic and Muslim values and also the promotion of real economic development. The latter appears as an absolute necessity to give some faith in the future to millions of men and women. The lack of a hopeful perspective could lead people to lock themselves in a spiritual environment to escape the material and moral ghetto in which they are confined by those in power.

To sum up, French public opinion is divided between two conceptions of French foreign policy in the Mediterranean. First, for French people who have no particular bond with the southern Mediterranean area, the policy should be essentially defensive. This view prefers France to endorse strong regimes capable of stopping immigration and Islam. Secondly, for French people who are sensitive to Maghrebi values, the policy should be to promote democratic transition and economic development. They would also favour a more liberal visa policy. Nevertheless the weak political weight of this community in the French political landscape and its incapacity to communicate through the media, mean a quasi absence of influence of their views in the elaboration of France's Mediterranean policy.

Because of the need to be profitable, the media rarely go against what they believe to be the prevailing views of public opinion. The following items have received the most media attention, and thus have been the most marketable issues in the French media since the early 1990s:

- a fear of expansionist and aggressive Islam;
- a tide of economic migrants;
- an extension of a 'second Algerian war' and its series of blind acts of terrorism;
- a flood of refugees escaping from a 'Khomenist' regime in Algiers;
- the increasing export of drugs from North Africa.

Sensational images, often deliberately taken out of their original context, as seen on the major television channels (mainly private channels such as TF1), feed and amplify the vague fear that the French have towards southern Mediterranean states. This process contributes to the spread of a siege mentality in France.

The consequences of this situation for the elaboration of foreign policy are well known: governmental authorities are sometimes compelled to follow the tide of the media to show that they understand their electorate. One should not underestimate the media pressure in the formulation of foreign policy towards the Mediterranean. Media pressure can provoke, in some sensitive cases, a change in French diplomacy. In October 1995 this was so with the Chirac–Zeroual meeting. Within ten days, this meeting, which had initially been presented as a "working meeting", became a subject of national controversy among the French political class.

The media, some influential and politically committed academics, and parties on the Left criticised President Chirac for endorsing the presidential candidature of General Zeroual and interfering in Algeria's internal affairs. With this media campaign President Chirac was compelled to announce that he would ask Zeroual to promise guarantees of democratic reforms for Algeria. Considering the well-known sensitivity of Algerian leaders, it was hardly surprising that the meeting was later cancelled by General Zeroual. The latter was able to retrieve a tremendous amount of sympathy in his country. Algerian voters were very proud to see the French President given a resounding refusal in full view of the international community.

In conclusion, French conceptions of what their Mediterranean policy should be converge on *the necessity to ensure the stability of the southern Mediterranean states*. The refusal of Islamism, the fear of a migratory wave, the anguish of becoming the target of blind terrorism, and the protection of economic and financial interests are the main reasons that motivate an endorsement of the established political leaders in southern Mediterranean states. Nevertheless the excesses of some North African leaders, notably in their lack of democratic values and violation of human rights, have led some sections of French public opinion and some high level officials to argue that French foreign policy should dissociate itself from, or at least put conditions on French assistance to these southern 'hard-line' governments. The relations between France and its neighbours in the Maghreb are still not neutral. The concept of having a common history is strong on both sides of the Mediterranean. For many French people, Mediterranean policy must contribute to ensuring that their country retains its traditional influence in this area.

## FRENCH FOREIGN POLICY IN THE MEDITERRANEAN AREA: A GRAND PROJECT ON A EUROPEAN SCALE OR AN ARBITRATION BETWEEN NATIONAL CONCEPTIONS IN EUROPEAN CLOTHES?

### François Mitterrand: the architect of France's Mediterranean policy in a European framework?

From a historical perspective, it is important to recognise the absence of a specifically Mediterranean policy in the past. In May 1981, when Mitterrand became President, France had no fully fledged Mediterranean foreign policy. Indeed, under the Presidency of Giscard d'Estaing, Paris had a juxtaposition of different policies in the Mediterranean area.[13] These policies were hardly coordinated and, more often than not, divergent:

- a European policy oriented to the South of the European Economic Community;
- a Maghreb policy based on passionate and difficult relationships;
- an Arab policy inherited from General de Gaulle that was torn between the will to keep a regional balance and the will to recognise the right of existence of the state of Israel and equally the right to a homeland for the Palestinian people.

*Mediterranean issues in the views of President Mitterrand*

For the new President at the Elysée in 1981, France, by virtue of its history, geography, politics and culture, would be the best placed country to build and promote a Mediterranean Community. The early 1980s marked for the Socialists the beginning of a new era. At that time the Mediterranean issues at stake were as follows:

*Diplomacy.* The stability of the Maghreb was a real preoccupation at that time when the Eastern bloc had become as intractable as ever, when daily violence occurred in the Middle East, when the Iran–Iraq war was at its height and when Afghanistan had been taken under Soviet control.

*Migration.* The forecast for a doubling of the Maghreb population by 2025 threatened a pressure on France's capacity to receive more immigrants. This possible immigration could not be ignored by the political authorities whose duty it was to secure France's internal stability. The high rate of unemployment and the need for structural adaptation which was demonstrated by the disappearance of several French industries such as coal, steel, and ship building, reduced France's welcome for foreign workers and more particularly for those from the Maghreb, in contrast with the situation during the period of post-war economic expansion called the "trente glorieuses". In the course of his first Presidency from 1981–8, President Mitterrand referred to the fact that the French population had nearly reached its threshold of tolerance concerning immigration.

*Trade*. Markets in the Mediterranean represented profitable outlets for French firms at a time when national commercial deficits with developed countries substantially increased and worried financial leaders who were already prone to disparage the Socialist experience and know-how. When French leaders saw, on the one hand, the decrease in their share of the western European market, and, on the other hand, the end of the possibility of so- called 'deals of the century' with Saudi Arabia and other Persian Gulf states, the Mediterranean appeared as an advantageous zone to be protected at all costs.

*Energy*. In the first half of the 1980s, agreements between the national company Gaz de France and Algeria's company Sonatrach made the southern shores of the Mediterranean a major and long-term source of energy supplies. Gas agreements are generally negotiated for several decades and involve heavy investment, notably in the construction of pipelines, large refinery units, and other infrastructure.

*The Mediterranean area: a major advantage for France in the European geopolitical game*

Just as the colonial empire had been presented by General de Gaulle during World War II as a symbol of the power of Free France ('la France libre') in his efforts to be recognised by the Americans and the British, the Mediterranean area constituted a key element of Mitterrand's European ambition. Indeed, in his concept of the role of Paris in European construction,

President Mitterrand was convinced that a strengthening of the European Economic Community's southern flank would be favourable to France and would increase the political influence and financial weight of the southern European states. This orientation aimed to place France in a privileged position to coordinate an increasing 'southern power' to counterbalance the growing influence of Germany. France is anxious not to fall behind the Federal Republic of Germany whose economic power continues to grow and is a matter of near obsession for French leaders.

Discreet initiatives in the service of a coordinated and progressive policy characterised the first efforts made by the Mitterrand administration. His long ministerial experience in the course of the Fourth Republic and his thorough knowledge of Maghrebi affairs gave him a distinct advantage over his predecessor Giscard d'Estaing, whose inexperience and lack of a long political vision had appreciably reduced the influence of France on the international stage. The 'man of 10 May 1981'[14] was aware that relations between France and its former North African colonies, particularly Algeria, were marked by emotion, sensitivity and incomprehension. For President Mitterrand these relations had to be managed in a more balanced way. This problem did not deter him from elaborating a plan for cooperation between neighbouring states in the western Mediterranean in which the preponderance of Paris could not be challenged. It was clear that the wider framework which was intended to bring a greater neutrality in North–South relations was accepted by French leaders with one condition: France should remain the main actor or in co-leadership if necessary.

In the Mediterranean area, the architect-strategist Mitterrand could also rely on two important allies: González, his Socialist friend who became Prime Minister of Spain in 1982, and Chadli Bendjedid, the Algerian President he met for the first time at the Cancun Summit in 1981. The difficulty of the exercise encouraged Mitterrand, on the one hand, to put in place in the Mediterranean a concerted and progressive policy and, on the other hand, to play the European card.

There were two emerging axes of French diplomatic action: the strengthening of French positions and influence in the western Mediterranean area and the development of France's locomotive role in European political construction.

Why concentrate on the western Mediterranean? Arguing that too rapid an extension of the geographical field would contribute to a dilution of dialogue and cooperation in the international field, Mitterrand asked his colleagues to promote a more limited framework, namely, the western Mediterranean area, so as to reduce contentious issues and remedy the weakness of the North–South dialogue.

Hubert Védrine, advisor to the President and later spokesman of the Elysée, considered that 'the western zone of the Mediterranean [is] a real totality in which common interests and problems are really interrelated'.[15] This concern of Védrine to provide a rational explanation for a choice essentially justified by strategic considerations is deeply anchored in French political culture.

On 27 January 1983, when addressing the Chamber of Representatives of Morocco, an audience that was not initially favourable and still less committed to him, the French President proposed to increase the opportunities for meetings between France and its 'immediate neighbours' in the Mediterranean. This proposal aiming to strengthen relations between France and the southern Mediterranean was not exclusive as Mitterrand, a declared partisan of European construction, spoke to his Moroccan hosts of extending these meetings 'to other countries to the north, south, east and west of the Mediterranean area'.[16]

This project of cooperation between states bordering the western Mediterranean was to be prepared and administered by two men: Védrine from the Elysée and Huntzinger from the Quai d'Orsay.

Social difficulties on the French domestic stage and the priority granted to European construction delayed President Mitterrand's Mediterranean projects. It was only in August 1987 (in the period of his cohabitation with right-wing Prime Minister Chirac) that the President revealed his intention to contribute to 'strengthening cooperation between Mediterranean countries and especially those in the western Mediterranean area'.[17] His chosen means was the *Mediterranean Forum*. The object of this initiative was 'to bring together, in an informal, open, and non-governmental framework, but still benefiting from the attention of the states concerned, a number of civil servants, politicians, socio-economic leaders, academics, and experts from European and Maghreb countries, to facilitate an exchange of views on, and a number of concrete proposals for, cooperation in the fields of economic, social and cultural affairs'. The Forum's objective was to 'facilitate and increase Mediterranean dialogue and cooperation between the societies and the states on its two shores.'[18]

The first Forum, presented as a private initiative, but supported at the highest political level, took place at Marseilles in May 1988. After this first meeting, a mission for the Mediterranean was established in the Quai d'Orsay, with ambassador Huntzinger being given charge of this dossier. According to one senior official, the depoliticisation of such an initiative was intended either to limit criticism of the Elysée if the project failed or, if it were a success, to treat it as an achievement of the presidency.

The European dimension constituted the second feature of President Mitterrand's strategy in the formulation of his Mediterranean policy. Concerned to ensure the Mediterranean anchorage of the Community, he insisted during the European Council meeting in Strasbourg that the EC deepen its neighbourly relations with the southern Mediterranean states. He argued that Europe had with those countries 'for a long time preferential links in the form of tariff agreements, and cooperation funds. This specific relationship with Mediterranean countries needs to be intensified to sustain them in their efforts towards cooperation with Europe, regional integration and economic development'.[19]

The Mediterranean policy of President Mitterrand apparently marked a break with the Gaullist tradition of independence concerning foreign policy especially towards Arab and African states. By proposing to widen French preferential relationships with former colonies in the Maghreb, President Mitterrand and his close advisers such as Védrine and Huntzinger, seemed prepared to share with European partners the privileged, even exclusive, relationships that France had maintained with these countries.

Behind statements and apparent intentions, the real conception of French experts on the Mediterranean dossier was that multilateral bonds would concern essentially financial and commercial links. Relationships between the EC and North African countries were essentially limited to the commercial field to the extent that European initiatives aimed to sustain regional development by increasing commercial exchanges. This increase rested on an opening of the European market and on the granting of reductions of customs duties. The financial dimension, in accordance with the commercial dimension, involved giving grant aid and loans for the long term to favour structural adaptation within the North African states, notably in response to the requirements of international trade. Under the influence of France, the Matutes Report, adopted by the European Council and the European Commission during the autumn of 1989,[20] argued for a reorientation of European actions and initiatives in favour of the Mediterranean area. This reorientation required an increase in the financial effort made by the EU based on the principle of regional co-development. The financial protocols increased from 615 to 815 million ECU. Nevertheless, European financial efforts in favour of the Maghreb still appeared insufficient if one bore in mind, on the one hand, what the South really needed in terms of investment and, on the other hand, the weakness of internal savings and the extent of the external debt. Recognising this situation and the limitations of their financial margin of manoeuvre, French officials tried to promote cooperation between public actors, notably at the European level, and private actors, essentially at the French level.

The adoption of the Matutes Report and the elaboration, in 1988–90 at the beginning of Mitterrand's second term as President, of a Renovated Mediterranean Policy marked a success for Paris in favour of a more important European investment in the Mediterranean area, where France alone no longer had the financial means to ensure stability.

Briefly, one can argue that during the fourteen years of Mitterrand's presidency the policy of France in the Mediterranean was articulated around two principles: the refusal of globalisation, and the necessity of a gradual and concerted approach with European partners.

At the first Mediterranean Forum, Védrine declared: 'We are well placed to know that because of history, all types of divisions and antagonisms are concentrated around the Mediterranean, be they real or imagined, involving religious, ethnic, and political clashes and a potential North–South confrontation'.

At the United Nations General Assembly, even during the Gulf War, President Mitterrand declared 'any global approach is unrealistic, and therefore harmful'. Possible cooperation between Europe and the southern states of the Mediterranean from Morocco to Turkey could be envisaged only as the last stage of a far longer process. It was with this perspective in mind that France proposed a 'dialogue à neuf' (a group of nine partners). Only seven states were present in 1989 during the second Forum: France, Italy, Spain, Portugal, Morocco, Algeria and Tunisia. During a European foreign ministers' meeting in Rome on 10 October 1990, the French succeeded in bringing partners to accept this idea of a limitation to nine participants in the western Mediterranean dialogue. On 18 October 1990 the French embassy in Rome informed Paris 'We have emphasised and made prevail . . . the specificity and the value of a dialogue with a framework of nine members in the western Mediterranean area.'[21] Nevertheless this French approach did not achieve the expected unanimity among European partners. Some of them, northern states notably, considered this proposal as an attempt by Paris to impose its pre-eminence in the western Mediterranean area.

The forum with nine partners was considered by the Quai d'Orsay as a first stage of a process aiming to insert French initiatives into a distinctly European framework. Within this framework, French diplomacy aimed to have a leading role so as to preserve a dynamic role in a traditional zone of influence. The plan was classical: Paris would open the economic dimension to its European partners while the political cooperation between Europe and its southern neighbours would be confined to a small number of French-led projects. The objective of the Mitterrand strategy was, on the one hand, to engage the EC in financing Maghreb development and, on the other hand, to preserve preferential bilateral relationships throughout the Mediterranean area. It is significant that the Italians, the Germans, and, more recently, the British partners, have received substantial business benefits from the Maghreb.

Finally, another even more discreet aim of President Mitterrand was to use this southern European dimension to constitute a French-led Latin group of four states within the EU.

## 1993: the return of the Right to power

The continuation of Mitterrand's tactics by the Right represented the consecration of a French European ambition. In the words of de Charette, the foreign minister, 'France is without any doubt the state that has to play a central role in the emergence of a Mediterranean dimension for the European Union. Other states can contribute, too, but I believe that France has a specific mission in that respect.'[22]

The multilateral dimension remained a priority in the Mediterranean foreign policy of France. Even more than European construction, the stabilisation of the Mediterranean appeared to be a priority for French diplomacy. Since the early 1990s, French diplomacy has launched great and expensive international initiatives with only hypothetical bases. They are characterised by both a carefully planned policy at the highest level, and preparatory meetings with small ad hoc groups made up of experts from the Grandes Ecoles such as the Ecole Nationale d'Administration.

The Mediterranean policy of France seeks to limit negative trends in a situation which is leading to a degradation of good Euro-Mediterranean neighbourly relations. There are different problems in each southern Mediterranean state. But each of them faces the same challenges: great demographic pressure; a weak social and educational infrastructure; high unemployment; insufficient diversification of production and industrial trade; weak regional trade; an inefficient and excessively expensive public sector; no future prospects for a large part of the population; a violent opposition to the ruling political class.

French Mediterranean policy is based on a new spirit and new instruments, and it aims to develop them in a coherent way within a clearly European framework. The Mediterranean policy can by summed up as having as its main characteristics a clear purpose, a reaffirmed European spirit, a strong French dynamic and a substantial content.

Four key words increasingly characterise the French diplomatic approach in the Mediterranean area: pragmatism, rigour, prudence and transparency.

French diplomacy is careful not to suggest to the southern populations of the Mediterranean that its actions in general and its involvement in the Euro-Mediterranean dialogue in particular are aimed to oppose political Islam and values on which it is currently based. Concerning the resolution of possible frontier disputes and minority problems, France considers that the mechanisms put in place on the European continent are not fully applicable to the southern Mediterranean area. Paris proposes an approach that takes account of Arab and Muslim cultures and of the need to treat the Mediterranean area as an autonomous zone.

The transparency of foreign policy, in both its formulation and its application, is a new characteristic sought by the French authorities and particularly by President Chirac. In the winter of 1995–6, the Foreign Minister insisted on the need to promote transparency in the Mediterranean policy of France. He declared, 'there was a time when diplomacy could be conducted with political cant. It was the period where diplomacy was practised by diplomats and statesmen alone. Today a significant part of diplomatic life is in the public domain. This is not a disadvantage, it is progress. This progress obliges us to speak clearly, to be understood, to talk in terms that public opinion can understand.'[23] The search for transparency had to develop a new credibility.

Since 1993 French foreign policy in the Mediterranean area has moved resolutely into a European dimension. The Foreign Ministers, notably Juppé (1993–5) and de Charette (1995–7), devoted great efforts to giving France a more dynamic Mediterranean policy. The declaration by de Charette to the Deputies in the Palais Bourbon (21 November 1995) was revealing in this respect:

> It is true that following a series of events, we have been kept apart from Mediterranean history. We have the intention to return to it and to act with determination. There is no common point between what we called the Mediterranean policy in the 1960s and the situation in the 1990s.

> Things have changed – there is no doubt; the environment has been deeply modified by a certain number of factors. Of course for France there is the European dimension, and you would not expect me to say, in one way or another, that I could ever renounce it. But it is not exclusive and it does not have to distance us from this field of concern which is extremely important, where very old friendships and very new problems are in evidence. For us it is an essential priority.

When Minister de Charette stressed that France had been kept apart, any observer would easily understand his reference to the Peace Process between Israelis and Palestinians. France and the EU were left out, on the one hand, from the Oslo secret negotiations in 1993 and, on the other hand, from the preparation of the historic signature of the Washington Accords, which took place under the sponsorship of Clinton and Yeltsin. This exclusion was felt as a humiliation. At the end of 1995, the situation changed. De Charette stressed the importance of this change:

> After Yitzak Rabin's funeral, a moving and tragic occasion, the President and [I] had a long meeting with Shimon Peres. Israel, that has long considered that its fate was not in the hands of Europe, is turning henceforth to us [Europe], and particularly to France, in order to prepare the new situation that the peace process between Israel and the Palestinians represents in that region. French foreign policy in the Mediterranean area is ambitious. This ambition has become one of our strongest priorities.[24]

This ambition would find expression in the preparations for the Barcelona Conference (27–8 November 1995). In the discourse of right-wing political leaders, France was presented as the country that had played a leading and decisive role to ensure the credibility of the Conference of Barcelona and of the Euro-Mediterranean partnership. Prime Minister Juppé, addressing by satellite representatives of Mediterranean and European states gathered in Malta for the Crans Montana forum, declared that in Barcelona 'the great innovation... is that we convinced the European

Union that it is not a contradiction to have a policy in favour of Eastern Europe and a Mediterranean vision'.[25]

The conclusions of the French Presidency issued at the European Council meeting held at Cannes underlined French diplomatic efforts in favour of the Mediterranean area.[26] Indeed the final document recorded that that Council took place in succession to those of Lisbon (June 1992), Corfu (June 1994), and Essen (December 1994). French diplomacy could rightfully claim the paternity of the historic Barcelona meeting.

On the financial and economic level, French diplomacy also underlined that it was under its own Presidency that Europe reached a favourable agreement for countries of the southern Mediterranean. Indeed it was under the French Presidency in 1995 during the Cannes meeting that the European Council decided to quadruple the budgetary assistance devoted to Mediterranean countries (4.7 million ECU). Thus, France expressed its solidarity towards the southern Mediterranean in a very substantial way, for all its neighbours, partners and southern clients to see. This demonstration of solidarity was achieved by involvement of the European Union which was intended to balance its interest in Central and Eastern Europe, an area considered to be the preferential zone of Germany with a strong interest in the Mediterranean.

In this context, the Mediterranean policy of France was close to being an arbitration policy in favour of the southern states:

> When one speaks of the arbitration that the European Union will have to make between credits devoted to Eastern and Central European countries and the efforts that must be made in the Mediterranean area, it is France and, without any doubt, France alone, that is able to propose good arbitration. Some states are attracted exclusively by Central and Eastern Europe, their fields of preoccupation and environment, others are attracted especially by the Mediterranean area, their neighbourhood. France is on both sides.[27]

French authorities have been able to rely on the French-speaking media, which has a particularly large audience throughout the Maghreb area. The objective of the Quai d'Orsay, through this means of communication, was to underline legitimately the efforts deployed by France to convince its

European partners to commit themselves to the Barcelona process. Not without reason these partners were also presented as initially reticent towards any important initiative in favour of southern Mediterranean states.

Aware of the criticisms generally directed at French foreign policy, such as hegemonic ambitions, arrogance and absence of dialogue, the French government committed itself to a real renovation of France's international image.

So, in order not to upset its Northern partners whose financial contribution to the process was required, French diplomats based in other European capitals rightly insisted that the Barcelona process reflected an EU commitment and not only a French one. This commitment focused on the consequences of the tremendous imbalance between the south and east Mediterranean states and those on the European continent. The main target of French diplomacy was to show EU northern member states that they too were concerned with the potential dangers in the southern Mediterranean. From such a perspective their financial contribution was fully justified.

Whilst working with their European partners, Prime Minister Juppé and Foreign Minister de Charette could also show that France did not use the European Union involvement as an excuse to decrease its own financial effort in favour of southern Mediterranean states. Indeed the cooperation policy of France with Mediterranean countries represented 1.2 billion FF in 1995–6.[28]

An example of such European and indeed Western solidarity was when France strongly rejected Colonel Gaddaffi's reconciliation proposals. This showed the will of President Chirac to give to French diplomacy a high degree of reliability.

For France the Barcelona Conference was only a stage in a longer process of promoting dialogue and free trade in the Mediterranean area. Dialogue and trade guarantee peace, prosperity, stability and the well-being of peoples who live on both sides of *mare nostrum* in the view of French political leaders and senior officials. The team that surrounded Juppé (when he directed the foreign ministry between 1993–5) was able to observe that, too often, Mediterranean policy had given rise to media opportunism without substantial results (cf. the Euro-Arab dialogue, the project of the Conference on Security and Cooperation of the Mediterranean or the Five-plus-Five initiative). Particular attention was given to the follow-up of the Barcelona Conference. For French diplomacy the Barcelona Conference and more exactly the specific initiatives launched there, were the object of a careful follow-up through ad hoc thematic meetings of ministers, senior officials and experts and through exchanges of views and information between public and private representatives.

The definition of three aspects (volets), 'politics and security', 'economics and finance', 'social and human affairs', gave a coherent and effective content to the Barcelona process. The financial package suggested by France and agreed by the European Union allowed it to implement the projects accorded to each of these three aspects.

The financial and economic assistance used mainly well-known instruments: aid protocols and debt postponements or cancellations. It was the intention of the Europeans that this assistance would help southern Mediterranean states to integrate with the world economic system. The Europeans hoped that this fund would support southern countries which were involved in a process of democratic reform. French political leaders and senior officials wished however to limit what they called a 'conditionnalité' in order to protect the southern Mediterranean leaders. French analysts underlined that this conciliatory behaviour was beneficial to Europe as a whole. Too strong a pressure on the southern Mediterranean leaders could accelerate a deterioration of their internal situation. A brutal confrontation with European economic competition would face southern political leaders with very unpopular choices concerning industrial restructuring and suppression of social benefits. French experts believed that too many unpopular decisions within a short period of time would favour, very probably, Islamic opposition groups.

Furthermore migratory movements could not be stopped in an environment of economic restructuring. Migration and the way to control it occupied an important place in France's Mediterranean policy. De Charette underlined this preoccupation: 'It is indispensable that, in the political debate that will be instituted with our southern partners, we help them understand that the future does not consist in crossing the Mediterranean Sea.'

French analysis asserted that the interest of European Union members was to favour stability, especially on the social level, and a peaceful transition to democracy. The actions of the Europeans should not serve the strategy of movements opposed to democratic values, even indirectly or involuntarily. Thus French officials considered that the prospect of free trade did not constitute a miracle solution. This perspective constituted only a necessary reform to accelerate integration into international trade.

The foreign policy of France within its European framework consisted of concrete proposals in favour of political dialogue, preventive diplomacy and common security. During the general debate which took place in France on this subject, the French Foreign Minister expressed the wish that principles which had been reaffirmed in the Barcelona Declaration's political aspect 'will take substance and life through a more concrete and operational instrument'. De Charette proposed the idea of a 'Mediterranean pact' or a 'Euro-Mediterranean pact'. Through this pact, the French authorities wished to formalise and especially to deepen the dialogue started in Barcelona. This deepening would make reference to the state of law and fundamental liberties but would avoid appearing as an attempt to interfere in the internal affairs of southern states.

By seeking to avoid interference, the French authorities indicated their refusal to reduce France's political action in the Mediterranean to a strictly ideological approach. The latter would not take account of the complexities and particularities of the Arab and Muslim world and would only emphasise both internal cleavages in the southern countries and misunderstandings between the north and the south of the *mare nostrum*. Generally speaking, but with certain differences of nuance, France's European partners were aware of the validity of the analysis made by French diplomats about the necessity of a progressive and flexible approach to the southern Mediterranean. For instance, Rifkind, then British Foreign Secretary, had supported the idea of a stability pact.[29]

The clear ambition of French diplomacy was to create the best conditions for a reconciliation between the two shores of the Mediterranean Sea. The main originality of the French approach consisted in an important financial and economic cooperation. This French approach distinguished itself from past initiatives that had privileged security issues. It was interesting that the new generation of diplomats working for the Quai d'Orsay was particularly sensitive to 'audit' methods. This awareness led them to examine past initiatives and, with a growing objectivity, to define reasons for failure or success. One could assume that the failure of the Conference on Security and Cooperation in the Mediterranean (CSCM), mainly supported by Italy and Spain was carefully analysed by the young experts of the Quai d' Orsay.

The promotion of security and stability in the Mediterranean area rested on a common appreciation or, at least, on a convergence by all states of perceptions of ways to promote stability, to share operational responsibility, and to contribute financially.

French diplomacy was seeking, through the dialogue which was a necessary but not sufficient means, to promote confidence between the countries North and South of the Mediterranean. This confidence was necessary in order to obtain the best common approaches possible in the management of Mediterranean security.

The creation of a Euro-Mediterranean College for Security Studies facilitated reconciliation and dialogue in the Mediterranean. This reconciliation depended on the development of a common Euro-Mediterranean awareness among senior officials, military officers and representatives of civil society who had a special vocation to serve in the Mediterranean area. Following the same general training and meeting to discuss common risks, various southern and northern partners would learn to compare their experiences and their methods of work, to communicate effectively, to comprehend more easily the specific needs of each of them when they are confronted with security questions, and above all to initiate relationships leading to mutual confidence.[30]

The development of common military exercises might lead southern Mediterranean public opinions to look at Europe in a better light. After the Gulf War, southern public opinions felt that Europe was refusing all relations that would have placed their governments on an equal footing. The creation of Euromarfor and Eurofor had been perceived from the southern viewpoint as an act of hostility or at least of distrust towards the southern Mediterranean area.[31] For a time southern public opinions thought that these forces were designed to operate in crises on the southern shores of the Mediterranean Sea. Common military exercises subsequently had as their main objectives to overcome this perception and to create understanding and confidence between North and South.

A number of concrete cooperation measures could equally concentrate on the following points: the fight against terrorism, humanitarian action, the protection of the marine environment, sea rescue, and the installation of a Euro-Mediterranean centre that would coordinate the supervision of this area and possible joint operations. Through these measures the Mediterranean policy of France was acquiring 'a new face': economy and efficiency constituted a real priority. It was noticeable that French and German analyses converged. German Foreign Minister Kinkel repeated at least three times in the course of a Barcelona press conference what Germany's wish for the Barcelona Process was going to be: "no new institutions, no new institutions, no new institutions".

Within a European framework, the pact proposed by de Charette constituted a priority for French foreign policy. In a first stage, this pact showed to its Mediterranean and European partners that France was still able to take the initiative on the international scene. This French initiative, proposed with the agreement of its European partners is independent from the United States. According to French analysts, this feature would help to convince southern states of their advantage in taking part in this new Euro-Mediterranean Partnership. This large-scale cooperation should facilitate dialogue and the management of sensitive issues through a mechanism involving small groups meeting regularly.

The Mediterranean pact could also constitute a neutral framework for the regulation of possible differences between states. It was possible that the foreign policy of France could find a European linkage to moderate difficult relations between particularly important countries for French interests, e.g., Morocco, Algeria, Tunisia, and Libya.

## FUTURE LESSONS AND PERSPECTIVES

Whatever the orientation of the political regime in France, its Mediterranean policy responded to the concern of maintaining preferential political relationships in an area of traditional influence. The vision expressed by President Chirac when he arrived at the Elysée in May 1995 did not distance itself radically from that of his Socialist predecessor. Indeed the will to promote a world role and responsibility for France largely transcended political cleavages.

If the objective stated by political authorities – the stability of the *mare nostrum* – has not changed in the course of the last two decades, the means available and the international geopolitical environment have changed considerably. France no longer has the necessary financial latitude to assume alone its role and its responsibilities in the Mediterranean. Europe appears to be a full partner nowadays. With the renewal of diplomats and officials one observes the development of a more professional approach. There is a beginning of a break with traditions, or at least habits, that gave an irritating and often a negative image of French foreign policy in the Mediterranean.

Regarding the future, in their joint letter to the European Council meeting in Madrid in December 1995, President Chirac and Chancellor Kohl called for the Union to have 'a common foreign and security policy which is more visible and more determined, in a strategy that better insures the efficiency, the continuity, the coherence and the solidarity of its action'.

In the wake of these declarations of intent, the European tax payer would not only expect greater synergy between France and Germany, the engines of European construction, but also with other Europeans, even with those who are farther away from the Mediterranean area, in order to promote the development of a European identity in foreign and security affairs. The United Kingdom for example, although it does not seem to have a clear policy on North Africa, should not be left behind by the other leading European nations. Firms such as British Petroleum and British Gas are already actively engaged in the Mediterranean and do possess some influence in the region.

Germany could bring benefits to Europe from its preferential links with Eastern Europe, and France could share its Mediterranean advantages with its neighbours who are undertaking already a part of the financial cost for the stabilisation of the area. The convinced European might distinguish here the embryo of a common foreign policy for the EU.

The success of the Barcelona agreement in November 1995 will be evident if the follow-up is substantial, if policies go beyond mere declarations of intent, and if it leads to a comprehensive convergence of foreign policies in theory and practice.

**Editorial Note:**

Following early elections on 25 May and 1 June 1997, a government of the Left was formed with Lionel Jospin, prime minister, Hubert Védrine, foreign minister, and Alain Richard, defence minister, providing Socialist Party leadership. But the President from the Right, Jacques Chirac, still controlled foreign and defence policies which did not change in the Mediterranean area.

## NOTES AND REFERENCES

1. For a detailed analysis of Arab policy during the 1960s see Jean Touchard, *Le Gaullisme*, Seuil, Paris, 1978.
2. For a detailed presentation of these messages see L. Mered, J. Williams, 'Arabes et Chirac', in *Stratex analysis*, August 1995.
3. Gilles Perrault, *Notre Ami le Roi*, Gallimard, Paris, 1992.
4. 1974–81.
5. 1981–95.
6. The Hotel Matignon is the office of the Prime Minister.
7. The RPR is the political party of President Chirac.
8. Michel Roussin is a former 'gendarme' and was the principal assistant of Alexandre de Marenches, when he was director of the French secret services. Michel Roussin was a member of Balladur's government from 1993 to 1994.
9. The French foreign ministry is known as the 'Quai d'Orsay'.
10. Algerian sources reported that there was a military committee in each newspaper to control press work.

11. A number of police operations against Islamic support groups connected to criminality in Paris, Lyon, Lille and Nice shed light on this threat.
12. Morocco, Algeria, Mauritania, Tunisia, Libya.
13. Valéry Giscard d'Estaing was President from 1974 to 1981.
14. Mitterrand was elected to his first Presidency on that day.
15. Speech of Hubert Védrine. Source: – CRAG – Université de Paris VIII.
16. Speech of the President, Chambre des Représentants, Rabat, Morocco, 27.1.1983.
17. Interview with President Mitterrand's advisers.
18. Jacques Huntzinger documents.
19. European Council of Strasbourg, Conclusions of the Presidency, 8–9 December1989.
20. The Matutes Report (autumn 1989) which was adopted by the Council and the Commission was an attempt to develop a new Mediterranean policy with greater financial support.
21. Quoted by CEDSI University, 1992.
22. Declaration of Hervé de Charette, then Foreign Minister, to the National Assembly, 21 November 1995.
23. Speech by Hervé de Charette, December 1995.
24. Declaration of Hervé de Charette, then Foreign minister, to the National Assembly, 21 November 1995.
25. Declaration of Alain Juppé, Friday, 13 October 1995, Crans Montana Forum.
26. Conclusions presented on 27 June 1995. Source: Presidency of the French Republic.
27. Declaration of the then Foreign Minister, 21 November 1995.
28. This figure is an approximation. Source: Foreign Business Ministry, December 1995.
29. Malcolm Rifkind gave his support to the proposal of Hervé de Charette during the Conference of Barcelona, on 28 November 1995.
30. For a more detailed description of the College see Mered, FMES Conference paper, June 1995 (Toulon) and CIDOB Conference on NATO, September 1995 (Barcelona).
31. These groups associate France, Italy, Spain, and Portugal.

# 3 Italy and the Mediterranean in the 1990s

Roberto Aliboni

After acquiring a high profile in the 1980s, Italy's policy towards the Mediterranean and Middle Eastern areas has undoubtedly been the subject of profound reconsideration in the subsequent decade, concerning both its goals and directions and its relative importance with respect to other areas. This change has been generated by external as well as domestic developments: the changes ushered in by the end of the Cold War and the situation ensuing from the longstanding domestic mismanagement of public finances.

Because of its financial difficulties in the 1990s Italy is finding itself increasingly deprived of the instruments and resources with which to conduct a foreign policy matching the international status that the country gradually acquired after post-World War II reconstruction. Furthermore, such difficulties have emerged at a time when relations within Western and European alliances are tending to become more competitive and less protective, thus preventing Italy from taking advantage of the combination of low-cost membership in the alliances and high international prestige deriving from that very membership as was the case during the Cold War.

For these reasons, Italy is in the throes of a difficult political transition, implying both constitutional and financial changes and the need to adapt its foreign policy. The outcome of this transition cannot yet be clearly identified. On the whole, however, a remarkable continuity seems to characterise both current developments and the future profile of Italy's foreign policy.

After a summary of Italian Mediterranean policies in the 1980s, this chapter will take into consideration the new tendencies that have emerged in the 1990s. Its conclusions deal with possible future prospects.

## REINFORCING MEDITERRANEAN POLICIES IN THE 1980S

It is a commonplace to point out that Italy's foreign policy during the Cold War was almost solely conducted along the lines and within the framework of the Atlantic Alliance and the European Community. Feeling weak sometimes with respect to the other members of the European Community, Italy tended to compensate for this weakness by staying close to the US in the Atlantic Alliance. But NATO's importance to Italy went far beyond its concern with the inter-European balance of power. Because of its geopolitical situation, Italy was able to obtain from the Alliance, and 'consume' for its national security, a good deal more than it ever could 'produce' and provide for the sake of the Alliance itself.[1]

Broadly regarded both by other European countries (particularly France) and by domestic opposition parties (Left and Right) as one of the staunchest supporters of the Atlantic Alliance and the US, Italy nevertheless retained a margin of independence in its foreign policy. In a recent article, Holmes very aptly says that this independence was the inherent product of an objective situation:[2] Italy was perfectly conscious of the fact that it would always be in the US interest to provide for the country's 'ultimate security', and so was the US. It was for this reason that Italy was able to develop a number of independent foreign policies and interests, especially towards adjoining areas like the Mediterranean and Eastern European.

But such independence has never really affected Italy's fundamental faithfulness to, and consistency with, the Alliance, nor has it brought about significant clashes between national and multilateral interests. If one refers to the Mediterranean, there were only two occasions in the course of the Cold War in which Italy's more or less well understood national interests threatened to turn this margin of independence into a clash between national assertiveness, on one hand, and the Alliance and the US on the other. The first was with the so-called 'neo-Atlanticist' approach developed in the 1950s by more leftwing leaders of the Christian Democratic Party (DC) such as Gronchi and Fanfani.[3] After the failure of the 1956 Suez expedition and De Gaulle's rise to power in

France in 1958, Italy's Catholic leftists thought that by increasing their role in the Atlantic Alliance they could convince the US to grant Italy – a non-neocolonialist country enjoying good relations with Arabs – a special role in the Mediterranean. While obviously nothing came of this approach, there were diplomatic and political tensions both domestically and in trans-Atlantic relations. The second occasion, a more straightforward clash, occurred during the *Achille Lauro*–Sigonella crisis[4], against the background of the assertive national Mediterranean policy developed by Mr Craxi and the attempt of the then Socialist-led government to elaborate a regional security policy in which Italy was expected to play a prominent role.

These two attempts remain exceptions, however, in a trend in which Italy's policies towards the Mediterranean were essentially consistent with its multilateral commitment in the Alliance. Nor should the *Achille Lauro*–Sigonella sequence be attributed too much significance in a decade which, broadly speaking, was very successful in coupling strict Italian support for the Alliance with a higher Italian profile. The result was that, especially in the eighties, Italy came to offer a much greater national contribution than the rather passive role of security consumer which had prevailed in the past.

The Italian role in the Alliance has been criticized by a number of analysts of the neo-geopolitical school of thought,[5] which has emerged in Italy after the end of the Cold War for having led the country to disregard its national interests excessively: as they are wont to say, Italy completely "delegated" its foreign policy to the US and the alliances (especially NATO). Others have remarked that, on the contrary, sticking to multilateralism in the Western and European alliances during the Cold War was the best way to pursue Italy's national interests.[6] Actually, the question of how national interests and multilateralism can be reconciled is less a question of the past (when this was well achieved by adhering strictly to Western multilateralism) than a question of the current post-Cold War era, in which national interests seem to prevail and it is not yet clear how international cooperation can be strengthened and attained. Italy had a different problem with its Cold War foreign policy and its margin of independence: the question is not whether Italy was unable or unwilling to assert the right quantum of national interests in the alliances but whether it was able to respond over time to changes in the international situation and growing demands from the alliances for new burdens to be shared and additional resources to be contributed.

In the 1980s Italian foreign policy definitely made a remarkable effort to respond to newly emerging demands and objectives in the alliances. This proved particularly true with respect to events in the Mediterranean and the Middle East. There is no doubt that non-Christian Democratic Italian prime ministers, such as Spadolini and Craxi, who led the centre–left coalition governments in the second half of the 1980s, provided an unprecedented boost to Italian foreign policy and that the Mediterranean and the Middle East played a remarkable role in this framework of enhanced international initiative.

After a few contributions to international peace operations – like the one in the Sinai multinational forces – it was in the Mediterranean and the Middle East, with the 1979 intervention in Lebanon, that the Italian government started to contribute military forces in a more systematic way to the international and multinational peace-related missions set up by the UN, the WEU and other intergovernmental groupings to intervene in the area.

Interventions in the Mediterranean and the Middle East were, at one and the same time, a driving force and the result of a debate[7] initiated in the same years on the need to give more weight to Italy's defence efforts towards the Alliance's Mediterranean southern flank with respect to the focus traditionally kept on the southeastern section of the Alliance's central front. The reshuffling of the main national defence missions envisaged by the Italian Chiefs of Staff, advocated in 1985 in a White Paper commissioned by Spadolini, began to give way to a change in Italy's military model entailing increases in forces' mobility, armaments and professionalism. The debate on the renewal of the Italian military model is still underway and, after a bold start, has not delivered very much so far. But it cannot be overlooked that the contributions to international intervention carried out by Italy, from Lebanon to Bosnia, have been made possible by the debate initiated in the 1980s in relation to the Mediterranean and the Middle East.

## COMBINING MULTILATERALISM AND BILATERALISM

These moves on the multilateral side were coupled by the strengthening of bilateral policies towards the areas south of Western Europe. There was a substantial increase in bilateral development aid. Bilateral political relations with the most important countries on the southern shore of the Mediterranean were expanded and upgraded.

Between 1980 and 1990 Italy's official development aid (ODA) increased from US $683 to $2,615 million at current prices (from $1,043 to $2,764 at constant 1989 prices), the most important increase ever in Italy's development cooperation resources.[8] In the same period, while sub-Saharan Africa continued to enjoy absolute priority as the destination of Italian aid, the Mediterranean and Latin America competed for second and third place. Regardless of priorities and percentages, the sheer amount of disbursements permitted by the overall aid increase turned aid to the Mediterranean into an important support for Italy's foreign policy (and economic penetration) in that area.

As for political relations, reference has to be made to two main developments. First, Italy's decision in 1980 to unilaterally guarantee Malta's neutral status and provide regular financial support to the island. Though Malta represented a minor risk on the whole, the Italian move was very congenial to the Alliance in view of the ambiguities of Dom Mintoff's government with respect to both Libya and the Soviet Union. Second, bilateral political relations and consultations with North African and Middle Eastern countries received an unprecedented boost by all leaders, from Andreotti through to Spadolini and Craxi. Inspired by the latter, the Socialists were the ones who envisaged the most articulated political project. The upgrading of relations with the most important countries of the Southern Mediterranean was aimed at establishing a kind of regional group with the task of securing stability and security in the area.[9] The group was intended to act in tune with non-Mediterranean interests and powers, especially the US, but without being overly dependent on them. At that time, there was much talk in Socialist quarters about the Contadora Group as a blueprint to be taken up and developed in the Mediterranean. Much of the course Craxi decided to pursue during the *Achille Lauro*–Sigonella crisis – in particular, the strong understanding the Italian government wished to maintain with Egypt – can be explained as an attempt to reinforce the grouping policy that was so central to Craxi's political strategy towards the Mediterranean.

Such a reinforcement of bilateral ties in the area made it possible for De Michelis, who became foreign minister in 1989, to envisage a foreign policy characterised by a particularly proactive and dynamic combination of bilateral policies inspired by national interests and multilateralism. Drawing on the experience Italy had accumulated in the 1980s, he tried to make the country something more than a staunch and passive – sometime even oblique – supporter of the Western and European alliances by bringing stronger Italian commitments and more constructive contributions into the framework of these alliances. While De Michelis' lifestyle and flamboyancy drew criticism, impatience and sarcastic remarks, in the end, his policy failed because of the unexpected changes in both international relations and Italian domestic affairs brought about by the end of the Cold War. Nonetheless, it epitomized fairly well how much ground Italy had covered by the end of the 1980s in moving from a sheer consumer of security to a more responsible partnership within the alliances.

Under De Michelis' guidance, Italy contributed to promoting two main initiatives in the Mediterranean: the Western Mediterranean Group, which was set up in Rome in October 1990, and the Conference on Security and Cooperation in the Mediterranean (CSCM), which was presented by Italy and Spain in Palma de Mallorca in September 1990 on the occasion of a meeting of the CSCE.[10] These new Mediterranean initiatives cannot be appreciated unless they are regarded in the wider context of De Michelis' foreign policy. In this context, Italy also started to promote a sub-regional understanding in Central and Southeastern Europe. These two regional directions of Italian foreign policy were expected to reflect national Italian interests and, at the same time, to reinforce European cohesion. While France, Spain and Italy were expected to lead European initiatives in the Mediterranean, Italy was supposed to cooperate with Germany towards Eastern Europe. For this policy to succeed, however, an essential factor was the strengthening of the European Union's new Common Foreign and Security Policy (CFSP). For this reason, Italy made a strong effort to promote the CFSP and to find a working compromise between Germany and France, on the one hand, and the United Kingdom, on the other, in view of the signing of the Maastricht Treaty. Indeed, for Italian regional initiatives to succeed and meet the national interests involved, a strong European Union has to be in place.

The efforts and projects just mentioned marked the apogee of Italy's Mediterranean policy. Thereafter, the changes brought about by the end of the Cold War and the Gulf War triggered the swift decline of this policy and made it come to nothing. Inherently weak, the CFSP was unable to stand up to the dilemmas and divisions generated in the European Union and the Atlantic Alliance by the break-up of Yugoslavia. Events in Eastern Europe and the unification of Germany created divisions and raised fears among the members of the Union and, along with developments in the former Yugoslavia, cancelled out the Italian initiative towards Central and Southeastern Europe. Finally, whereas the Western Mediterranean Group came to a standstill because of the sanctions on Libya and the beginning of the crisis in Algeria, the CSCM and the potential for a European initiative towards southern areas were overwhelmed by the Gulf War and subsequently replaced by the US role in the Madrid process.

In conclusion, somewhat ironically, Italy's Mediterranean and Middle Eastern policy at the end of the 1980s suffered a collapse at the very time when it appeared to have ultimately reached a fair and consistent combination between multilateralism and national interest in the Mediterranean as well as in the wider circle of its foreign policy.

## EMPHASISING MULTILATERALISM IN THE 1990S

After the end of East–West confrontation, all Western countries were faced with the need to adapt their foreign policies to the emerging international environment. But this task proved particularly difficult for Italy because the end of the Cold War also brought an end to the bonus provided to Italy by its special geostrategic position in the Alliance. In the more demanding and competitive relations characterising post-Cold War multilateralism within Western and European alliances, Italy was suddenly confronted with higher costs than its allies. Furthermore, while facing such additional costs, Italy also had to withstand the consequences of the financial mismanagement of the 1980s and the collapse of its by-now eroded domestic political system (a system in many ways related to the Cold War situation and Italy's role in it). Thus, the international weakness was coupled with a particularly difficult domestic situation.[11]

The parliamentary elections of Spring 1996 brought in a government supported by a left-wing coalition. If this coalition managed to maintain its cohesion while pushing through political reforms and harsh financial restructuring, Italy might achieve the transition initiated in the Spring of 1992 with the collapse of the last Andreotti cabinet under the weight of appalling corruption and such flagrant financial mismanagement as to put into question Italy's participation in the common European effort to establish a streamlined monetary and economic union pursuant to the Treaty of Maastricht.

In the few years since the major domestic crisis in 1992, the transition towards a more stable political system was at first characterised by remarkable (even by Italian standards) instability: Andreotti's last government was followed by five cabinets. Governments changed every year, from June 1992, when Amato was appointed as prime minister, through to 18 May 1996, when the current government of Prodi was sworn in.[12] It must also be said that, unlike what used to happen in the political system of the Cold War era, that instability had a more systemic character, as it worked in a political arena that had not managed to find the new shared institutions it required (for example, an effective new electoral system). Furthermore, the right-wing government led by Berlusconi attempted to introduce a system of winner-take-all by changing leadership not only at the top of the government hierarchy – as was the case in the previous system – but also at intermediate and lower levels of power. This attempt stirred shock waves in the country and, unsuccessful and short-lived as it was, contributed to further instability and turmoil.

All these factors hardly eased the steering of national foreign (and defence) policies in the 1990s. Moreover, they hardly allowed these policies to be in the forefront of the Italian political stage. Intertwined with the financial and economic restructuring of the country, EU and other European policies have become the almost all-absorbing focus of both domestic and foreign policy. But, while European policy tends to be turned into a dimension of Italian domestic policy, foreign policy as a whole – including Mediterranean policy – has been neglected and sacrificed. Foreign policy cannot but reflect the sharp and repeated reductions that have taken place since 1992 in the state budget and affect such diverse and important policy instruments as international aid or military forces for peace-related missions. In order to fund the small contingent sent to Bosnia, the government had to introduce an *ad hoc* additional tax on gasoline. As for development aid, this decade has witnessed a continuous reduction of available funds: bilateral ODA went down from 3,354 billion lire in 1990 to 1,110 in 1994 and this trend has continued.[13] In recent years, foreign policy has often been reduced to a sort of declaratory policy, strongly dependent for real action on the policies carried out within multilateral alliances.

This dependence on multilateral policies must be regarded as the outstanding characteristic of Italy's foreign policy in the 1990s. In fact, whatever the balance-sheet at the end of the decade, there is no doubt that Italy's foreign policy in the 1990s will be noted for seeking to resist tendencies towards re-nationalisation, to restore and strengthen multilateralism and upgrade its ability to play multilateral games as the sole way to meet national interests and maintain or regain its international prestige and status.

This trend may appear somehow ironical in view of the fact that, with the collapse of the political class that ruled during the Cold War, a very vocal public debate has been raised by the emerging neo-geopolitical Italian school of thought about the necessity to work out a foreign policy predicated on the assertion of national interests (as these interests – so the argument runs – were neglected during the Cold War) in a world that is going to be dominated once again by competition and confrontation among nations. It was thought that the advent of Berlusconi's government in May 1994 would give way to such an assertive policy, but it soon became apparent that Italy was unable to assert anything at all and that its policy of national assertiveness in Istria as well as in the European Union did nothing but isolate and weaken it internationally without any appreciable result. The end of the Berlusconi government only eight months after it was sworn in cut short any illusions or expectations of national assertiveness and fully restored Italy's traditional multilateralist approach. As a consequence, any extremism incurred by the neo-geopolitical school of thought was quickly cut down to size. Today, very few would doubt that Italy's primary interest is in strong multilateralism and the debate refers mostly to diverse emphases or options in an essentially multilateral framework.

Italy's interest in the Mediterranean has, however, been strongly challenged by the trend referred to here as the Italian neo-geopolitical school. Two main arguments have been developed by this school: first, that the Mediterranean is no more than a second-rank option within the range of Italy's national interests (and that Italy's geopolitical focus has to become the European East as well as the global economy);[14] second, that, to the extent to which an interest in the Mediterranean survives, Italy should concentrate on a few countries, like Egypt and Libya, rather than dwell on comprehensive schemes of regional or inter-regional relations.[15] In relation to these countries, Italy should not be interested in complying with allied policies seeking to distance themselves from the policies of the regimes in power towards political Islam or to isolate rogue countries by international sanctions and so forth. A broader argument is that since the interest of the most important EU members in the Mediterranean is marginal, concentrating Italy's efforts on this area would not pay off in terms of European influence and status.

The need to turn more decisively from the Mediterranean and the Middle East towards Eastern-Central and Southeastern Europe is voiced with particular vigour. Italian investment in Eastern Europe, particularly in Romania and Albania, and economic relations with Russia and Southeastern Europe have actually significantly increased since the end of the Cold War. But the persistent weakness of the

Central-European Initiative (the so-called Pentagonale) should remind Italy of its fundamental lack of national political instruments in the region. Despite the very good prospects for Italian economic penetration in Eastern Europe, Italian interest in the development of an area like the Mediterranean, from which its gas and oil essentially come, cannot be lower than its interest in Eastern regions. The argument could be, rather, that an increasing interest in the East is only natural with the return of this area to co-operative international relations. All in all, Eastern Europe and the Mediterranean were always important directions for Italy during the Cold War and will continue to be so in the future.

Whatever the substance of these arguments, the Italian governments of the 1990s – despite some confused attempts by Berlusconi's cabinet to work out a more 'national-interest-based' approach – have lent a deaf ear to this debate and, rather more in tune with both traditional Italian policy and rhetoric, have pursued a Mediterranean policy which reflects the broad weaknesses of the country in its current painful transition. The major weakness lies in Italy's inability to pursue a bilateral policy as strong as that carried out at the end of the 1980s. In fact, as already noted, the thrust of Italy's Mediterranean policy in the 1990s is to be found almost exclusively in the multilateral sphere. True, the modesty of bilateral contributions weakens Italy's presence and role within today's Mediterranean multilateral policy frameworks, from the Euro-Mediterranean Partnership to NATO's Mediterranean Dialogue. Nevertheless, Italy's efforts in these frameworks should not be underestimated both *per se* and as a factor that may well upgrade Italy's role in the alliances as soon as the domestic transition is succesfully completed and a stronger bilateral contribution is once again possible.

The debate about the Mediterranean is not limited, however, to that between the (mild) nationalist tendencies of the neo-geopolitical intellectual circles – a selective bilateral approach, mostly directed towards the European East – and the more balanced multilateralism of the governmental circles. To complete the picture and obtain a better definition of the Mediterranean's role in the Italian society of the 1990s, other tendencies are worth mentioning.

Despite continuity in Italy's mainstream international policies and perceptions, a set of political and economic domestic factors related to the Mediterranean area have as a matter of fact changed in the wake of the important transformations brought about internationally and domestically by the end of the Cold War's global regime. Such changes do not bring into question the importance and the role of the Mediterranean component in Italy's foreign policy and national security as the neo-geopolitical school does – nor do they oppose governmental policies. Nonetheless, they contribute to shifts in visions, emphases and relative importance with respect to the past that may be harbingers of a more articulated and finely tuned Mediterranean policy for this country.

In this sense, one can identify four new principal trends in Italy's attitudes and thinking with repect to its neighbours in the South.

First, the civil society that is emerging from the collapse of international communism and the subsequent profound transformations in Italy's leftist parties, trade unions and associations, has lost the anti-imperialist and non-aligned rationale that underpinned policies towards the Mediterranean. The strong solidarity and support provided by the Italian left to trends such as Arab nationalism or the support of the Palestinian struggle have faded, or in the case of Palestine, have assumed a different perspective. With the exception of the old-fashioned Communists gathering in the small Party for the Restructuring of Communism, Italy's Left today is certainly more attracted by the European and trans-Atlantic circle and looks at the Mediterranean from a standard European vantagepoint, both politically and ideologically.

This trend in the former Communist Left (which today is the most important support for the governmental coalition) does not mean that the Mediterranean is neglected in Italian foreign policy. Rather it suggests that, with respect to the exceedingly important role that the Mediterranean was given in domestic politics and foreign policy rhetoric during the Cold War, the ranking of the Mediterranean in Italy's foreign policy is just bound to become more attuned with national realities and interests.

Second, the fragmentation of the former Christian Democratic party into various smaller Catholic parties, and the weakening of the cross-cutting Catholic–Communist stream that undercut most of Italian politics during the Cold War era are both factors that contribute to a radical change in the attitudes of Italy's civil society towards the Mediterranean. The Catholic–Communist stream used to promote a set of cultural and political myths of Mediterranean solidarity which were then translated into anti-Western, anti-European and anti-American attitudes. As these myths and related policies were rooted in a search for authenticity, they were not dissimilar to the trends against modernity and the West promoted on the other side of the Mediterranean by the political movements inspired by Islam. This second change contributes to consolidate a more relaxed and realistic view of the Mediterranean and, therefore, coalesces with the new Europeanist look of the governmental Left in giving the Mediterranean a more realistic size in the context of the Italian interests.

Third, the two aforementioned trends are reinforced by the important changes undergone by the foreign policy of the Vatican with respect to the Third World in general and the Mediterranean in particular. Since the 1960s, but particularly with the the present non-Italian Pope, the foreign policy of the Catholic Church is increasingly directed by the fundamental interest of protecting Catholics and Catholic interests. This strategy is particularly evident in the Middle East and the Mediterranean and recently culminated with the establishment of normal diplomatic relations between Israel and the Vatican. Such a move would have been hardly accepted by the strong Arabophile (and anti-Israel) posture of the left wing of the former Christian Democratic party and the Catholic–Communist streams of opinion. On the contrary, the move was broadly welcomed. This fact is evidence that the combination of anti-imperialism and anti-modernism that sustained Italy's Mediterranean vision in the post-World War II era is now dissipating.

Fourth, the swift changes introduced in international economic relations by strong trends to globalisation and relative modifications in the distribution of economic power among old and new industrial countries are affecting the importance of the Mediterranean in Italy's economic interest abroad. Obviously, as already pointed out, the opening up of the

Eastern markets has attracted Italian products and capital. Even more attractive to Italian businessmen are Asian markets. However, the relative decrease of the Mediterranean's share in Italy's economic relations does not correspond to a reduced importance of this area for the Italian economy. Energy, for example, remains a primary factor in Italy's relations with the Mediterranean. Despite the huge efforts of diversification made by the national oil company in the last two decades, the Mediterranean's importance in Italy's energy supply remains very high. The recent project for importing Libyan gas will add to Italy's interdependence with the Mediterranean. Even a limited success of the free trade area envisaged by the Euro-Mediterranean Partnership will be highly beneficial to Italy and will upgrade again the relative importance of this area in Italy's external relations. While there are a number of political and ideological changes which tend to adjust the Mediterranean's role in Italy's foreign relations to a more realistic role, economic interests continue to assign to this area a remarkable importance.

The combination of a durable economic importance with a more pragmatic and sober political role in Italy's Mediterranean policy may ultimately contribute to dismantle the Mediterranean rhetoric accumulated during the Cold War and make Italy's Mediterranean policy more effective. This perspective is consistent with current multilateralism. Still it may give substance to the necessity of working out a reinforced bilateral policy.

While these new forces and fresh trends are working to shape Italy's Mediterranean positions, perceptions and policies in the short as well as the middle term, we have to take into consideration the concrete policy conducted so far by the Italian government in the 1990s if we want to give a more complete picture.

## THE ITALIAN GOVERNMENT AND THE MEDITERRANEAN

All five governments of the post-Cold War era, including Berlusconi's, have more or less stressed their attachment to the Mediterranean direction in Italian foreign policy.

However, the most conscious and articulated supporter of Italy's need to promote a Mediterranean policy was Ciampi in his capacity as Prime Minister from April 1993 through May 1994. Ciampi's foreign policy was predicated on two strong assumptions: first, that the European nations have to remain firmly framed by the system of alliances established after the tragedy of World War II, even if there are prices to be paid for that (an assumption that continues to inspire him in his current task of Minister of the Treasury in Prodi's government); second, that Italy has a basic interest and role to play in the Mediterranean area, and the task of maintaining a link between the Mediterranean and the 'Mitteleuropean' worlds. In a speech given at the end of his mandate, he pointed out that the European Union's cohesion and success rest on its ability to "hold a fair balance internally between the two principal components of its identity: the Mitteleuropean and the Mediterranean. Both of them are equally vital and have, in the course of centuries, enlivened one another.'[16]

Ciampi promoted Italy's Mediterranean initiative in two directions: in the Mediterranean sphere itself, by helping to establish the Forum for Dialogue and Cooperation in the Mediterranean in cooperation with Egypt and a set of other countries belonging to the area; [17] and in NATO, by convincing the other members of the Alliance, in tandem with Spain, to affirm the necessity of a trans-Atlantic move towards the area.[18]

The Forum was an initiative of the Egyptian government, by which it intended to allay frustrations and concerns over the regional and international changes brought about by the end of the Cold War. At that time, Egypt began to perceive that the Arab–Israeli peace process, the implications of the 1990–91 Gulf War and the policies pursued by the US from its all-dominant position in the Middle East, were altering the strategic balance and threatening its traditional leadership in the region.[19] Consequently, Egypt was also concerned that the huge amount of foreign aid extended to it by both the US and the Gulf Arab countries – largely because of its important strategic role in the region – could be put in jeopardy in the less immediate future. The Mediterranean option emerged at the end of 1993 as an element of a new Egyptian strategy to check this decline in influence and was manifest in the proposal to establish the Forum as a link with the southern members of the EU as well as in Egypt's request to join the Arab Maghreb Union and become a member of the (sleeping) Group of Five Plus Five (i.e. in what must have appeared to Cairo as a sphere of privileged cooperation with Europe). Briefly, the emerging prospects in the region persuaded Egypt to promote Mediterranean solidarity as a further option in the region and as a means for strengthening ties with the European Union.

The idea was first launched in early 1993. In January 1994, the Egyptian initiative was favourably received by Ciampi and Italian Foreign Minister Andreatta. Throughout the first semester of 1994 the Italian diplomacy committed itself to developing the Forum and to this end asked the International Affairs Institute (IAI) in Rome to prepare a report on what could be done to implement cooperation in concrete terms within the Forum.[20] This report – entitled 'Med-2000' – was then adopted as the terms of reference of the working groups which were eventually established by the Alexandria ministerial meeting of the Forum in July 1994.

The meeting in Alexandria was held at a time when the first building blocks were being placed by the EU on the way towards the Euro-Mediterranean Partnership – EMP (finally formally established in Barcelona in November 1995). The EMP was very eagerly espoused and supported by Italy from the European Council in Corfu (June 1994) to the one in Cannes (June 1995), and that soon made the Forum obsolete. Although the Forum continued to meet and work, its identity had been greatly weakened by the fact that the Forum's principal aims were pursued far more effectively by the EMP. The question that can be asked in relation to these developments is why the Italian government pursued the Forum when the process that was to lead the EU to Barcelona had already been initiated.

The answer is that in the first semester of 1994, it was not very clear where the EU and the Mediterranean were going to go from there. There was a feeling among Southern European countries that a policy initiative was needed to relink Europe to the Middle East and the Mediterranean after the Gulf War and the course assumed by the Madrid process had weakened European trans-Mediterranean ties. Besides, there was also the feeling that action had to be taken to rebalance the unilateral eastward direction that the emerging European Union's CFSP was taking. In this framework the Italian diplomacy saw the Forum as an instrument with which to stop the drift of the Mediterranean with respect to Europe (and the drift of Southern Europe with respect to Northern Europe) and grasped it.

To be sure, the Forum could be regarded as a framework reproducing some elements of Craxi's past aspirations to set up some kind of Mediterranean caucus, but there is no doubt that the Italian government had no intention of creating a Mediterranean solidarity in opposition to a Euro-Mediterranean scheme. At that time, such a Euro-Mediterranean scheme was vaguely envisaged – in Italy and elsewhere – as an adaptation of either the longstanding EU Mediterranean policy or the idea of the CSCM (Conference for Security and Cooperation in the Mediterranean). In fact, the IAI's report, presented by the Italian government in Alexandria, included a proposal for setting up a Euro-Mediterranean form of political cooperation very similar to the one that is presently being implemented within the Barcelona process.

Subsequently, Italy co-initiated and very firmly supported the Barcelona process. During its EU presidency in the first semester of 1996, the Italian government took various initiatives particularly devoted to developing the cultural dialogue between Europe and the South Mediterranean partners in the EMP. Most importantly, Italy successfully steered the Senior Official Committee in its task of drawing up the blueprint for the 'Action Plan' on which the EMP's security policies are predicated.

In fact, there is strong political continuity between the Forum and the Italian contribution to the Barcelona process, in the sense that Italy is pursuing the same political goals in the EMP that it wanted to attain in the Forum. No doubt, Italy feels more at ease in the stronger multilateral framework provided by the EMP. True, the Forum was also a multilateral framework, but it could not take advantage of the means and the political authority now made available by the Euro-Mediterranean process started in Barcelona. For a country whose bilateral policies and instruments are constrained by domestic problems, the EMP is certainly providing a more effective framework than the Forum. Nonetheless, Italy has not dismissed the Forum. This may indicate that it continues to be attracted by the idea of a Mediterranean caucus besides the EMP, which could be revived in the future and find a role for itself in the narrower Mediterranean arena.

The second basic idea put forward by Ciampi that contributed to forming the backbone of Italy's Mediterranean policy in the 1990s is that NATO has to play a role in the relations with South Mediterranean countries. After the NATO Summit of January 1994, in which the idea was strongly promoted by the Italian government with French and Spanish support, Spain took up the initiative (at the NATO meeting in Seville in September 1994) and, at the beginning of 1995, the Mediterranean Dialogue became an official initiative of the Alliance, though it was inaugurated by a very unfortunate declaration by the Secretary-General, in which he gave the impression of equating political Islam to Communism as the new enemy of the Atlantic Alliance. The bad impression created was never really dispelled in subsequent developments.[21]

The Mediterranean Dialogue – like the WEU Dialogue – has not made any significant progress since its inception. NATO officials are very careful to point out that the Dialogue is nothing more than what the word suggests and that it has no other task than that of providing mutual information and correcting southern Mediterranean misperceptions about NATO and its transformation after the end of the Cold War.[22] However, this respectable task is something less than what its initiators had in mind. Such a limited approach reflects the fact that NATO has in principle accepted a Mediterranean dimension to its activities but is still not really prepared to develop it for the time being. An enlargement to include Eastern European countries, the intervention in Bosnia, and the establishment of a European identity within the Alliance seem to be more urgent matters on the NATO agenda than Mediterranean initiative.

Nonetheless, Italy took up the initiative again. At the informal October 1995 NATO meeting in Williamsburg, the Italian Minister of Defence, General Domenico Corcione, urged his colleagues to consider the launching of a 'Partnership for the Mediterranean' (PfM), modelled on NATO's Partnership for Peace with Central European countries. In this and other statements on the same subject,[23] Gen. Corcione provided varying suggestions for the possible contents and tasks that could be given to the PfM, from 'soft security' and 'political counterproliferation' to policies to prevent the proliferation of weapons of mass destruction. He stressed the necessity of a gradual and prudent approach in an area that is very different from Eastern Europe, including the need to start with the most moderate countries on the southern shore of the Mediterranean.

There is no doubt that Italy's PfM initiative was developed with some American understanding. It is known that the initiative was discussed between General Corcione and US Secretary of Defense Perry before being disclosed. In any case, the available literature[24] reveals that a group at the Rand Corporation in Santa Monica was exploring the same topic. At the beginning of 1996, the Italian Military Centre for Strategic Studies (CeMiSS), part of the staff of the Ministry of Defence, initiated a research project on the PfM in cooperation with the Rand team and at the same time requested the IAI in Rome to conduct an enquiry on the same subject. This work was directed at enabling Italy, perhaps with US support, to table a more substantive proposal on NATO Mediterranean activities when the time was ripe for adding the need for progress in the Mediterranean to NATO's agenda.

Italy's EU- and NATO-related Mediterranean initiatives underlined the continuity of a kind of two-track policy that was not new in Italian post-World War II foreign policy. It shed light on Italy's lasting interest that the US be present in both the Mediterranean and the European arenas and, more broadly speaking, that the trans-Atlantic dimension merge with the European one. It also shed light on the fact that, as usual, the Italian military tended to emphasize the trans-Atlantic framework, whereas Italian diplomats were closer to the European sphere, though there was a strong convergence on the fact that both contexts must be present to secure a maximum of stability and security for the country.

This two-track policy was reflected in the policy conducted by Italy in the 1990s with respect to the inclusion of a security and defence component in the European Union. Italy has always supported the strengthening of a European security and defence identity, but on the condition that it be closely related to NATO. In the end, this was Italy's position in the negotiations of the Intergovernmental Conference which led to the signing of the Maastricht Treaty. In the same period, Italy avoided joining the French–German Eurocorps for fear of creating cleavages between Europe and the US. Today, Italy is perfectly at ease in strongly supporting the process that should bring about the drawing up of a European defence identity inside NATO. Meanwhile, the interest in NATO goes hand in hand with interest in European military integration. In fact, the two-track policy is also reflected in Italy's participation in the WEU's Euromarfor and Eurofor, in cooperation with France and Spain, for implementation of military instruments essentially devoted to performing Petersberg-like missions in the Mediterranean area.

## CONCLUSIONS AND PROSPECTS

As already pointed out, Italy's policy towards the Mediterranean in this decade is deeply affected by an adverse combination of international and domestic changes.

In this changing framework, there have been political and cultural trends towards a more assertive foreign policy, mostly directed at Eastern Europe, thus de-emphasising the traditional importance assigned by the country to the Mediterranean. These trends, however, have come to almost nothing. In fact, Italy's governments have continued to pursue significant Mediterranean policies and the accentuation of multilateralism as a way out of the numerous current constraints. Consequently, Italian initiatives have mostly emerged in European and trans-Atlantic multilateral spheres.

The inception of the Barcelona process has provided Italy with a multilateral framework in which it can conduct a Mediterranean policy more significant than that which it could otherwise carry out bilaterally or even in cooperation with the other Southern European nations. In addition to its inherent political and military relevance, the development of a Mediterranean dimension in NATO is above all a reassurance for Italy against the risks of becoming marginal on the

EU political stage as well as in the framework of the EU Mediterranean policy – hence the traditional two-track Mediterranean (and European) policy mainly predicated on multilateral spheres, which secures not only strong continuity but even accentuates Italy's past preference for multilateralism.

In perspective, this picture poses two questions. From the point of view of the European Union's political process, the adoption of the new EU Mediterranean policy in Barcelona amounts to closing the perceived gap between East and South and integrating Eastern and Southern policies in the EU Common Foreign and Security Policy. In other words, the Barcelona process means reinforcement of EU cohesion. The first aim for Italy is that its two-track policy should contribute to strengthening trans-Atlantic links without weakening recently reinforced European cohesion. This issue goes beyond the Mediterranean and does not concern Italy alone, for – in a sense – almost all European countries have a two-track policy and a more or less strong interest today in developing a European defence identity within NATO. However, frustrations and tensions in difficult intra-European negotiations might push Italy to emphasise trans-Atlantic reassurances instead of supporting European cohesion and its French–German core. The successful integration of Italy in the first circle of the EU monetary and economic union would reassure Italy and help it conduct its two-track policy fairly and cooperatively.

The second question, which pertains more closely to the Mediterranean, is that the weakness of the bilateral policy component cannot last for long. This weakness is not felt by Italy alone, but by the entire range of Southern European countries. The Mediterranean and Middle Eastern policy of Greece is almost solely dominated by its concerns with Turkey; Spain has not yet been able to devote sufficient resources and instruments to conducting a strong bilateral policy in addition to its remarkable contribution in the multilateral sphere; and domestic problems have compelled Italy to decrease sharply the resources for the Mediterranean, which had enjoyed a considerable increase in the 1980s. Upon election, President Chirac stressed the role France is willing to play in the Mediterranean, the Middle East and Africa, but the country continues to be absorbed by the need to reshape its European, trans-Atlantic and global role and, despite some initiatives, like the one in Lebanon in the Spring of 1996, it is clear that France is unable to play the same role south of Europe that Germany is currently playing in leading and combining multilateral and bilateral policy components to the east of the European Union.

Italy alone cannot play in the Mediterranean the role that Germany is playing in Eastern Europe, nor can it replace France in the area. It can, however, contribute to improving the situation. This means that Italy has to reinforce its bilateral policy towards the Mediterranean, as soon as this is possible, unless it wants to run the risk of jeopardising its multilateral efforts. Besides, Italy has to continue to strengthen its successful cooperation with France and Spain in an effort to introduce a strong Southern European element in the EU Mediterranean policy. In this sense, the establishment of Euromarfor and Eurofor[25] within the WEU is a right step both in the region and in the context of the European and trans-Atlantic alliances.

## NOTES AND REFERENCES

1. Stefano Silvestri, 'L'Italia: partner fedele ma di basso profilo', in Luigi Caligaris (ed.), *La Difesa europea: Proposte e sfide*, Istituto per gli Studi di Politica Internazionale, Edizioni di Comunità, Milan, 1990, pp. 185–202.
2. J. W. Holmes, 'Italy: In the Mediterranean, *but* of it? *Mediterranean Politics*, vol. 1, no. 2, 1996, pp. 176–92.
3. See: Sergio Romano, *Guida alla politica estera italiana*, Rizzoli, Milan, 1993, pp. 86 and 97–101; Paolo Cacace, *Venti anni di politica estera italiana (1943–1963)*, Bonacci, Rome, 1986, chapters XLII–XLIV.
4. See Gianpaolo Calchi Novati, 'The Case of the *Achille Lauro* Hijacking and Italo–Arab Relations: One Policy, Too Many Policies, No Policy', *Journal of Arab Affairs*, vol. 10, no. 2, 1991, pp. 153–79; Annette Jünemann, *Italiens Nahost Politik von 1980 bis 1990*, Nomos Verlagsgesellschaft, Baden Baden, 1993, chapter 3; Bijan Zarmandili, *Documenti di un dirottamento. Il caso 'Achille Lauro' nei giornali e in televisione*, ERI, Turin, 1988.

5. See the opening statement and, in particular, the contributions by Ernesto Galli Della Loggia, Angelo Panebianco and Gian Enrico Rusconi in the first issue of *Limes. Rivista di geopolitica*, no. 1–2, 1993, pp. 7–26. Also S. Romano, op. cit., and Ernesto Galli Della Loggia, 'La morte della patria. La crisi dell'idea dinazione dopo la seconda guerra mondiale', in Giovanni Spadolini (ed.), *Nazione e nazionalità in Italia. Dall'alba del secolo ai nostri giorni*, Laterza, Bari, 1994, pp. 125–61; Carlo M. Santoro, *La politica estera di una media potenza. L'Italia dall'unità ad oggi*, Il Mulino, Bologna, 1991.
6. Cesare Merlini, 'Cinque punti sul nazionalismo italiano', *Il Mulino. Europa* vol. 2, no. 43, 1994, pp. 89–99.
7. C. M. Santoro (ed.), *L'elmo di Scipio. Studi sul modello di difesa italiano*, Istituto per gli Studi di Politica Internazionale, Il Mulino, Bologna, 1992; S. Silvestri (ed.), *Nuove concezioni del modello difensivo italiano*, CeMiSS, ed. Rivista Militare, Rome, 1990; M. Cremasco (ed.), *Lo strumento militare italiano*, IAI, Franco Angeli, Milan, 1986.
8. Pierangelo Isernia, *La cooperazione allo sviluppo*, Il Mulino, Bologna, 1995, pp. 143–5 and 242–9.
9. See Roberto Aliboni, 'Institutionalizing Mediterranean Relations: Complementarity and Competition', *Internationale Politik und Gesellschaft* (Bonn), no. 3, pp. 290–9.
10. Ettore Greco and Laura Guazzone, 'Continuity and Change in Italy's Security Policy', in R. Aliboni (ed.), *Southern European Security in the 1990s*, Pinter, London and New York, 1992, pp. 69–85 ; Laura Guazzone, 'Italy and the Gulf Crisis', in N. Gnesotto and J. Roper (eds), *Western Europe and the Gulf*, WEU Institute for Security Studies, Paris, 1992, pp. 71–87.
11. See 'IAI Report 1992 – The Dual Crisis', *International Spectator*, vol. 27, no. 1, 1993, pp. 5–30.
12. Between the governments of Amato and Prodi there were those of Ciampi (sworn in 29 April, 1994), Berlusconi (11 May 1994), and Dini (17 January 1995).
13. José Luis Rhi-Sausi (ed.), *La crisi della cooperazione italiana, Rapporto CeSPI sull'aiuto pubblico allo sviluppo*, Edizioni associate, Rome, 1994, p. 11.
14. Carlo Jean, *Geopolitica*, Laterza, Bari, 1995, chapter 11 and p. 260; Ludovico Incisa di Camerana, 'I presupposti di una nuova politica estera italiana', *Relazioni Internazionali* (Milan), March 1993, pp. 62–70; Carlo Maria Santoro, 'Rischio Mediterraneo', *Relazioni Internazionali*, July 1994, pp. 2–3. For Italy's global-economy connection see Marcello Pacini,

'Geoeconomia, la parola chiave per orientarsi nel mercato futuro', *La Stampa* (Turin), 3 December 1995.

15. The journal of the new geopolitical Italian school, *Limes*, published an issue (No. 2, 1994) devoted to Italy's relations with the Mediterranean and the Middle East in which the editorial introductory article, 'Arabia Infelix', insisted on this point. The same issue contained an article published under the pseudonym Linus, underlining Italy's national interests in Libya: 'Perché ci serve Gheddafi', pp. 227–32.
16. Carlo Azeglio Ciampi, 'L'eclissi del bipolarismo e il nuovo multipolarismo instabile: riflessioni sulla ricomposizione del sistema internazionale a livello globale e regionale', in Istituto Affari Internazionali, *L'Italia nella politica internazionale. Anno ventunesimo*, SIPI, Rome, 1994, pp. 251–8.
17. Algeria, Egypt, France, Greece, Italy, Morocco, Portugal, Spain, Tunisia and Turkey, later joined by Malta. On this Forum, see Roberto Aliboni, 'Institutionalizing ...', op. cit., p. 297; *The Mediterranean Forum*, documenti IAI, IAI9512, Rome, 1995.
18. See point 22 in the press communiqué issued by the Council of the Heads of State and Government held in Brussels on 10–11 January 1994.
19. Abdel Monem Said Aly, 'From Geo-Politics to Geo-Economics. Egyptian National Security Perceptions', in UNIDIR, *National Threat Perceptions in the Middle East*, Research Paper no. 37, United Nations, New York and Geneva, 1995, pp. 17–30; Mohammed El-Sayed Selim, *Mediterraneanism: A New Dimension in Egypt's Foreign Policy*, Al-Ahram Centre for Political and Strategic Studies, Strategic Papers no. 27, March 1995.
20. See: 'Cooperation and Stability in the Mediterranean: An agenda for Partnership', *International Spectator*, vol. XXIX, no. 3, 1994, pp. 5–20.
21. G. W. Winrow, 'A Threat from the South? NATO and the Mediterranean', in *Mediterranean Politics*, vol.1, no.1, 1996, pp 43–59.
22. Sergio Balanzino, 'Security Challenges in the Mediterranean Region', in R. Aliboni, G. Joffé and T. Niblock, *Security Challenges in the Mediterranean Region*, Frank Cass, London, pp. 187–92. (Balanzino is Deputy Secretary-General of NATO).
23. The Williamsburg statement is summarised in *Atlantic News*, no. 2735, 6 October 1995. A subsequent statement was made in Turin at the 41st Annual Session of the North Atlantic Assembly on 9 October 1995 and can be found in the official proceedings of the Assembly.

24. F. Stephen Larrabee and Carla Thornson, *Mediterranean Security: New Issues and Challenges, Conference Proceedings, Brussels, October 15–17, 1995*, Rand, Santa Monica (CA), 1996; Ronald D. Asmus, F. Stephen Larrabee and Ian O. Lesser, 'Mediterranean Security: New Challenges, New Tasks', *NATO Review*, May 1996, pp. 25–9.
25. See details on Euromarfor and Eurofor in John W. Holmes, op. cit.

# 4 Spain and the Mediterranean

Carlos Echeverría Jesús

## INTRODUCTION

Spain has dramatically changed during the past twenty years. In 1986 it joined the European Communities (EC) which was an event of tremendous symbolic importance. In the same year the country voted to remain in NATO. Spain is now undergoing a process of Europeanisation and the country's most immediate task is to eliminate the gap that exists with most of its European partners. At the same time, as a southern European country it cannot ignore the Mediterranean for reasons connected to its own regional interests.

Spain is one of the few European countries to be both Mediterranean and Atlantic-oriented in terms of language and culture. Its traditional Jewish and Arab influence also impact on its foreign affairs. For instance, the traditional 'special relationships' of Spain with the Arab world have always been central to its foreign policy and a number of active policies toward several Mediterranean countries (especially in the Maghreb) have been developed. Nevertheless, Spain has had very little consciousness of the Mediterranean as a 'global region'. In fact, as such, the Mediterranean itself has figured only peripherically in Spanish foreign policy.

More recently, southern Europe and Europe as a whole have become more concerned with a number of issues and problems emanating from the Mediterranean. In turn these, broadly defined, now figure prominently in Spanish diplomacy.

Europe is heavily dependent on the Middle East and North Africa for its energy supplies, as nearly 60 per cent of its needs in hydrocarbon supplies come from there. In 1993, Algeria and Libya supplied 19.4 per cent of Europe's requirements in natural gas, with a further 6 per cent to be supplied by the new Algeria–Spain gas pipeline via Morocco which was completed in 1996, bringing the total up to one quarter of Europe's natural gas import needs[1]. This new pipeline links the Algerian fields with Spain and, from there on, to the rest of Europe.[2] In the years to come the oil and gas pipeline itineraries will determine the national security policies of the supplier, the supplied and the transit states as well.

A Maghreb that is shaken by political instability, economic crises and religious extremism would make Spain into the inhospitable frontier of a prosperous Europe. With increasing prosperity Spain has emerged as a significant destination as well as a conduit for migration. Radical Islamism and illegal migration constitute a barometer of the degree of distance that has been allowed to develop between the two opposing shores of the Mediterranean. In political terms, the western European planners are making support for the status quo in North Africa a central, if fragile, plank of their Mediterranean policy. This is seen as essential to reduce the threat of what European capitals fear to be a new generation of radicalised, anti-western states emerging on their southern flank. The immigration pressure has long been a divisive issue in France with its large North African population. More recently, it has emerged as a potent political question in Spain and Italy, where more tolerant attitudes toward immigration have traditionally prevailed.[3]

## DOMESTIC SOURCES OF SPANISH FOREIGN POLICY

The role of King Juan Carlos I is particularly significant in Spain's relations with the non-European Mediterranean countries. His official visits – for example to Egypt in February 1997 – must be emphasised. The King is in charge of representing the Spanish state in international relations (Article 56 of the 1978 Constitution). The representative figure of the King of Spain in international relations has become very important, so much so that through him the overall prestige of the country has grown.

In general, Prime Ministers ('Presidents of the Government') under the democratic regime – Adolfo Suárez (1976–80), Leopoldo Calvo Sotelo (1980–82), Felipe González (1982–96) and José María Aznar (1996–) – have taken a personal interest in foreign policy. The formulation of foreign policy by the government as a collective entity (Article 97) implies in fact that the policy is directed by the Prime Minister. In 1994–95 the role of Felipe González was decisive in persuading Chancellor Kohl of the necessity of rebalancing EU aid between eastern and southern neighbouring states.

The centre of foreign policy formulation and execution is the Ministry of Foreign Affairs.[4] The period 1990–95 was one of intense lobbying by the Spanish Foreign Ministry, headed first by Francisco Fernández Ordoñez and then by his successor, Javier Solana. In 1992 Fernández Ordoñez drafted a report on the political situation in the Maghreb commissioned by the EC foreign ministers in the context of the European Political Cooperation (EPC). In 1996, the appointment as Foreign Minister of Abel Matutes, the former EC Commissioner in charge of Mediterranean affairs,[5] underlined Spanish involvement in Mediterranean developments. In December 1995, Javier Solana was appointed Secretary-General of NATO.

The appointment in November 1996 of the Spanish Ambassador to Israel, Miguel Angel Moratinos, as special EU envoy to the Middle East underlined the importance of certain individuals in the shaping of Spanish foreign policy. The Spanish have been astute enough to extend its earlier focus on the Maghreb to the wider Middle East, also including Israel in their proposals. According to Pollack, 'Latin America and the Middle East are two areas where Spanish foreign policy has been at its most consistent'.[6] The special envoy will fulfil his mandate in close cooperation with all parties in the region and complementary to the important role which the US plays in the peace process.[7] Moratinos, one of the main architects of Spanish Mediterranean policy,[8] convinced Mauritania and Israel to move towards diplomatic relations during the Euro-Mediterranean Conference held in Barcelona in November 1995.

The Parliament[9] has become a major source of foreign policy not only in the framework of its Foreign Relations Committee but also in the form of parliamentary debates. The Parliament assures the control of governmental foreign policy (article 66) authorizing the ratification of certain international treaties. For instance, in November 1995, the Spanish

Congress of Deputies approved unanimously a motion that rejected the Association Agreement between Morocco and the EU because of Spanish economic interests in the fishing industry and the farming sector. The role of a number of Spanish parliamentarians who are also members of some international assemblies – the European Parliament, the WEU, NATO, OSCE and Council of Europe's Assemblies and the Interparliamentary Union – must also be mentioned.[10]

Certain regional entities – the so-called 'Comunidades Autónomas' – are playing a growing role in dealing with their 'foreign relations'. The Catalonian President, Jordi Pujol, or the Andalusian President, Manuel Chaves, have paid official visits to a number of North African countries where capital from Catalonia and Andalusia is being invested.[11] The fishing and agricultural communities of Andalusia and the Canary Islands are fearful of losing their livelihood through EU economic concessions to Morocco and other southern Mediterranean countries.

The most important Spanish lobbies or interest groups are concentrated in the energy sector. For instance, large investments have been made in the construction of the Maghreb – Europe gas pipeline from Algeria to Spain via Morocco. Gas Natural and Repsol are increasingly interested in signing agreements with Algeria, Egypt or Libya.

Dealing with public opinion in general, issues such as human rights in a number of southern and eastern Mediterranean countries, especially Algeria, Morocco, Israel/Palestine and the Western Sahara are a matter of public discussion in the media. Finally, the very modest Euro-Arab University in Granada and a number of research institutes and departments of Arabic Studies or International Relations at the university level are strongly engaged in work on Mediterranean issues.

## A 'MEDITERRANEAN POLICY' FOCUSED ON THE MAGHREB

In 1975 Spain was confronted with the 'Green March' organised by Morocco on the Western Sahara. Spain opted to abandon its responsibility for the territory by handing over its administration to Morocco and Mauritania. The Spanish government stated that the decolonization process would not be completed until the Sahara people had freely expressed their will.[12] The Saharan question put an end to Franco's 'policy of balance' based on exploiting the differences and rivalries between Morocco and Algeria so as to check Moroccan claims on Ceuta and Melilla.[13] What actually happened, however, was that Morocco and Algeria used Spain as an instrument in their disputes. In 1983 President Felipe González announced that Spanish foreign policy towards the Maghreb would be based on cooperation aimed at promoting stability and integration. It would not be a 'policy of balance' but a 'global policy' for the region.

During the 1980s it was thought that Spain's historical 'special relationship' with the Arab world might facilitate a more influential role for Spain in the western Mediterranean.[14] The admission of Greece (1981), and of Portugal and Spain (1986) to the EC also focused attention on Mediterranean issues. This particular enlargement was seen by some third Mediterranean countries as an important threat to their exports to the EC. In January 1986 full relations between Spain and Israel were established. Spain's position on the Middle East went beyond what the European Council Venice Declaration of June 1980 had called for, namely a 'comprehensive settlement'. Felipe González stated that 'a just and peaceful solution in the Middle East will have to be based on Israel's retreat from all of the Arab territories occupied since 1967'. The government formally recognized the PLO office in Madrid shortly thereafter.[15]

Spanish policy with respect to the Arab world picked up considerably after the end of the Cold War. Europe's displacement towards the East put Spain in a peripheral position. In the post-Cold War era, Spain was influenced by the impact of the Gulf War, and saw its interests more clearly tied in with the South, making the Maghreb a priority objective inasmuch as it was a risk area for Spain's global security.[16]

In fact, Spain has succeeded in consolidating bilateral relations, based on pragmatism that takes its inspiration from a major concern: the political and economic stability of the Maghreb. Spain's policy is based on channelling economic resources towards the region as a preventive measure in favour of Spanish security. The four-year Cooperation Agreement signed between Spain and Morocco in June 1988 – an amount of Pta125,000 million – is a good example of this new policy. In addition, in 1991 a Friendship Treaty was signed between the two countries.[17] Spain also signed with Algeria a soft loan of Pta135,000 million on 2 February 1989, for a three-year period. In October 1995 in Tunis, Spain signed a Friendship, Good-neighbourliness and Cooperation Agreement with Tunisia.[18] Also in military aspects Spain wants to create a web of relations with the Maghreb countries. The Spanish government signed in November 1987 a Military Cooperation Agreement with Tunisia. On 7 February 1989 Spain signed another similar Agreement with Mauritania. Military cooperation between Spain and Morocco includes reciprocal visits to military centres, and Moroccan cadets and officers attending Spanish military schools and academies; joint exercises are also conducted and overflights arranged. There exists an ample trade in arms and technical support.[19]

For the time being, practical Spanish interest in the Mediterranean is giving rise to new perspectives and new policy concerns. Tunisia and Morocco are the two most promising partners for investment in the Maghreb[20] and there appears to be growing interest in Algeria on the part of European oil and gas companies including Spain's *Repsol*[21]. Spanish private-sector investment in Morocco in already quite high and Tunisia is considered today as part of the new emerging markets. The middle-income countries of the southern Mediterranean offer new opportunities for investment, partnership and development, in which western Europe is set to take a leading role. Moreover the political, economic and cultural revitalisation of the Mediterranean countries would place Spain in a region of growing prosperity and at the centre of a rich cultural dialogue which could make the South into an important alternative for Spain in the future.

For the time being, a number of structural problems are threatening this optimistic scenario. With its pivotal position between the Mediterranean, the Atlantic and sub-Saharan Africa, Morocco has long been a major source for hashish, which is shipped across the straits of Gibraltar into Spain.

With the increasing mobility of populations from sub-Saharan Africa, Morocco is also becoming an important transit area for illegal immigration onto the Mediterranean's northern shore. Ceuta and Melilla are not part of the EU customs territory, and have a different trade status to the rest of the country, based on the respect for their traditional international free trade status. The Moroccan government is critical of the lack of progress achieved on the future status of these enclaves. In 1992 Morocco had agreed to the return of African illegal immigrants held in the Spanish enclaves; in 1996 for the first time since the signing of this agreement Rabat has agreed to re-admit illegal immigrants.

In July 1987 the last bilateral Fishing Agreement between Morocco and Spain expired. Morocco, who had applied to join the EC in July 1987, maintained very ambitious aspirations which were reflected in the negotiations on the new Agreement for fishing rights on its heavily used continental shelf. On 25 February 1988 a four-year Fishing Agreement was signed in Brussels between Morocco and the European Community. Morocco won some financial and technical advantages but the most important point was to allow the signing of a more favourable Financial Protocol and other protocols adapting the 1976 Cooperation Agreement between Morocco and the Community to the new twelve-state EC. For the time being, of the aproximately 700 fishing boats concerned, about 93 per cent are Spanish.[22] On 15 November 1995 a new four-year Fishing Agreement and an Association Agreement with the EU were concluded.[23] It reinforces Morocco not only economically but also politically, including it within the Euro-Mediterranean partnership framework announced at the end of November 1995 at the Barcelona Conference. In fact, the new Fishing Agreement with Morocco was one of the prerequisites for success at the Barcelona Conference.

The Spanish government seeks to forge cooperation rather than seek confrontation, and its policy towards the North African countries includes non-interference in internal affairs. The arrival of the first Maghrebi 'boat people' in Spain has given impetus to this policy – even when helping existing North African regimes means sometimes overlooking human rights violations. For instance, although European governments expressed concern at the human rights situation in Algeria, there was tacit support in much of Europe and the West for the High Council of State takeover in Algiers as the only possible way to avert the rule of the Islamic Salvation Front (FIS). The question of democracy is crucial to the future of any working Euro-Mediterranean relationship, especially in terms of achieving enduring security. Business and economic measures are useful but democratic developments are also essential.

This finds expression in efforts by EU governments to provide the maximum possible political and financial support for Algeria, Morocco and Tunisia, and for closer security cooperation. Spanish policy-makers believe that a great deal of good will come out of efforts by Spain, France, Italy and others states to shift southwards the political and economic attention of the EU. When the Arab Maghreb Union (UMA) was founded in Marrakesh on 17 February 1989 by Algeria, Libya, Mauritania, Morocco and Tunisia, the Spanish government was the first European state to react by supporting this politically relevant development.[24] The UMA represented a long-held aim to bring North African states, often divided by political rivalries, closer together, both economically and socially. It was considered an increasingly significant factor in the complex nexus of relations binding the northern and the southern flanks of the Mediterranean.

## SPAIN AND SECURITY AND COOPERATION IN THE MEDITERRANEAN

Spain has also reinvigorated its southern engagement through multilateral activities. For instance, there is a growing role of Spain in European initiatives aimed at development and stability in the Mediterranean region. In fact, the southern European states are adopting the role of bridge-builders between the southern and the northern shores of the Mediterranean Basin.

Efforts to engage non-European Mediterranean countries in a political and economic dialogue were given new urgency by the extent of popular support in the southern Mediterranean countries for Iraq during the Gulf crisis and war of

1990–91. This sensitivity provided much of the impetus for proposed or active initiatives from the 'global' concept of a Conference on Security and Cooperation in the Mediterranean (CSCM) to more limited exercises such as the 'Five Plus Five' grouping in the western Mediterranean, the Mediterranean Forum,[25] the EU's Renovated Mediterranean Policy or acting then within European Political Cooperation (EPC) as a lobby for converting the Maghreb into one of the areas for the new Joint Actions, the equally new Common Foreign and Security Policy (CFSP).

The initiative for a CSCM was first outlined by Spain at the Open Skies Conference in Ottawa in February 1990.[26] The CSCM proposal was presented on 24 September 1990 at the opening of the CSCE meeting on the Mediterranean at Palma de Mallorca.[27] The matter was also raised on 10 October 1990 in Rome at the Ministerial Meeting of the 'Five Plus Four' grouping. The 'Five Plus Four' process considered that the most pressing problems in the western and central Mediterranean were political and economic in character. The CSCM sought to be a global answer to Mediterranean security challenges by adopting a set of generally accepted rules and principles to be followed after the solution of crises and by dealing at the same time with aspects of security and disarmament, together with issues of economic cooperation or relating to the 'human dimension'.[28]

The outbreak of the war in the Gulf, however, had negative effects on the CSCM, and after the war, the US gave priority to the Middle East peace process launched at an international conference in Madrid in October-November 1991. Although the initiative for the CSCM was later re-adapted by France, Italy, Portugal and Spain, the implementation of this conference has been prevented mainly because of US opposition as well as developments in Algeria and Libya. In fact, there was very little follow up to this joint Spanish-Italian initiative which has subsequently been developed by the Inter-Parliamentary Union.[29]

The Mediterranean region was undergoing profound change as the EU was adjusting to a new political configuration. In December 1990 the European Council adopted the Renovated Mediterranean Policy, which defined its relations with the Mediterranean non-member countries. This Policy called for agreements that would replace the cooperation agreements that were due to expire in 1996, and that would include new features such as political dialogue with the southern partners.

In 1992, at least in the first instance, Spain pushed for pursuing cooperative measures in the western and central Mediterranean where the dominant issues are of a political and economic rather than of a military nature. The main policy objectives of the EC/EU in the new era were summarised in the European Council Lisbon Declaration of June 1992: Maghreb stability is important for all EU members; priority should be given for a constructive dialogue for the establishment of a zone of peace, security and prosperity; it will be necessary to build a comprehensive framework of cooperation.[30] Spanish foreign policy considers that these cooperative measures would be easier with the UMA as a whole. Nevertheless, relations between the Maghreb states are still affected by the disagreements which emerged after Morocco requested a temporary suspension of the UMA's institutions in November 1995 because of Algeria's position on developments in the Sahara dispute, and also after disagreements between Morocco and Tunisia surfaced because of other Maghreb issues. The violent conflict in Algeria is now the central political fact of the western Mediterranean and it affects the economic prospects of the entire region. Morocco looks more to Europe for the future. Two-thirds of its trade is already with the EU.

On 27–8 November 1995 the Euro-Mediterranean Conference was held in Barcelona with the participation of the fifteen member states of the EU, eleven states from the southern and eastern Mediterranean, and the Palestinian Authority. This Conference was one of the most important successes of the Spanish presidency of the Council of the EU. The ensuing Barcelona Declaration defines the broad outlines of a Euro-Mediterranean Partnership. The Conference addressed the political, economic and cultural dimensions of a Mediterranean dialogue. Felipe González steered the Spanish presidency to success at the Madrid European Council Meeting in December 1995, where EU leaders agreed to the Barcelona process.

North–South cooperation in the western Mediterranean still remains active, at least on some sectorial issues. For instance, on 19 April 1996 the Interior and Justice ministers from Algeria, France, Italy, Morocco, Portugal and Spain held the Conference of the western Mediterranean states on the fight against drug trafficking in Rabat. They agreed on developing cooperation among them in the economic, technical and judicial aspects. They also stressed the need to continue strengthening coordinated and controlled monitoring of the border areas in order to prevent the flow of drugs, arms or illegal immigrants across the Straits of Gibraltar.

Although political and economic cooperation has been the focus of these initiatives, all have longer-term security implications. Close collaboration with Italy and France on matters relating to security in the western Mediterranean has tied Spain closer into the web of bilateral agreements binding western European countries. Spain perceives dialogue with the southern Mediterranean countries on security and defence matters as the most valid mean to improve knowledge of each other and to remove mutual misperceptions and misunderstandings. For instance, during the 'Tramontana 94' military exercise involving France, Italy and Spain in November 1994, the Spanish government invited military attachés from the countries of North Africa to attend as observers.[31]

After its accession to the Western European Union (WEU) in 1988, Spain supported the development of a security dialogue with a number of non-WEU Mediterranean countries. Since 1992 the WEU has been gradually developing a dialogue with some countries of the Maghreb region (Algeria, Mauritania, Morocco and Tunisia) which was later expanded to Egypt in 1994 and Israel in 1995.[32] On 8 February 1995 exploratory discussions were undertaken between NATO and Egypt, Israel, Mauritania, Morocco, and Tunisia (expanded to Jordan in November 1995). The initial input was provided by Spain and Italy. In fact, the NATO dialogue was launched in response to a Spanish proposal.[33] King Juan Carlos I supported the continuation and improvement of the NATO's Mediterranean initiative during his official visit to the Atlantic Alliance's Headquarters in May 1996.[34]

## CONCLUSIONS

The strategy followed by Spain to defend its own interests in the Mediterranean region has deeply involved bilateral and multilateral instruments. Apart from purely geographic factor, Spain's relations with the Mediterranean region must be seen as a key element in viewing Spain as an international actor. In fact, with a succession of proposed or active initiatives, southern European countries in general, and Spain in particular, have found a new and far more active diplomatic role in the world.

Mediterranean countries must make every effort possible to avoid their basin remaining a troublesome area and attempts are being made to promote projects which can create sustainable economic growth as a way to support stability in the region. At the same time, every opportunity for dialogue must be utilised by Mediterraneans to extend the negotiating environment which exists in Europe to areas of the southern and eastern shores of the Mare Nostrum.

Spain is emerging as one of the most important actors in the evolution of Euro-Mediterranean relations. In the early 1990s Spain emerged as leading player at the regional level during the gestation of the EU's Euro-Mediterranean Partnership Initiative. The CSCM proposal and the 'Five Plus Five' grouping were useful vehicles for political activism both in a European and a North–South context. The WEU and NATO bilateral relations and their separate Mediterranean dialogues have been fully promoted by Spain. The main achievements for this key player include the November 1995 Euro – Mediterranean Conference, the starting point of the Barcelona Process in which many hopes have been placed. The Barcelona Conference was indeed the culmination of a process that the EU had initiated in the late 1980s with the aim of revising its policies toward the Mediterranean nonmember countries in light of major changes that were taking place on both shores of the Mediterranean. In fact, the definition of Mediterranean security in comprehensive terms reflects the perceived vulnerability of Europe to developments beyond its Mediterranean shore.

## NOTES AND REFERENCES

1. The other link between Algeria and Italy via Tunisia was inaugurated in 1984.
2. In 1996, Spain depended on Algeria for almost 70 per cent of its natural gas supply.
3. Because Spain still thinks of itself as a net exporter of labour there is no great public concern, as is the case in France, at imminent Maghrebi migration.
4. Foreign Affairs Ministers under the democratic regime: Marcelino Oreja Aguirre, José Pedro Pérez Llorca, Fernando Morán, Francisco Fernández Ordoñez, Javier Solana and Abel Matutes.
5. The so-called Renovated Mediterranean Policy was introduced by Commissioner Matutes in 1989–90. At the end of 1992, the responsibility for Mediterranean matters in Brussels passed from Abel Matutes to Manuel Marín, a Spanish Commission Vice-President.
6. See B. Pollack, *The Paradox of Spanish Foreign Policy. Spain's International Relations from Franco to Democracy*, London, Pinter, 1987, p. 81.
7. See IISS Strategic Comments: 'Europe and the Middle East peace process' London, IISS *Strategic Comments*, vol. 2, no. 10, 1996 p. 2, and *Summary of World Broadcasts – BBC* 14 November 1996, p.6.
8. The conceptual formulation of the CSCM proposal was largely set up by this senior Spanish diplomat.
9. The Cortes (the Spanish Parliament) is made up of two Chambers: the Congress of Deputies and the Senate.
10. e.g. the European Parliament veto to the Fourth Financial Protocol between the EU and Morocco in 1992.
11. See C. Echeverria Jesús, 'La presencia de las Comunidades Autónomas españolas en el Magreb', in B. Lopez García (ed.), *Atlas de la Immigración Magrebí en España*, Madrid, Ediciones UAM, 1996, pp. 210–12.
12. C. Alonso Zaldivar and M. Castells, *Spain beyond Myths*, Madrid, Alianza Editorial, 1992, pp. 215–16.
13. Spain has five enclaves in the northern Moroccan coast: two cities – Ceuta and Melilla, two rocks – Vélez de la Gomera and Alhucemas, and one archipelago, Chafarinas. Spain occupied Melilla in 1497 to protect its coasts from the pirates. Ceuta, conquered by Portugal in 1415, also became part of the Kingdom of Spain in 1580. Ceuta is home to over 73,000 Spanish nationals; Melilla to over 63,000 Spanish nationals.

14. See R. Gillespie,'Spain and the Maghreb' in R. Gillespie, F. Rodrigo and J. Story (eds), *Democratic Spain: Reshaping External Relations in a Changing World*, London, Routledge, 1995, p. 159.
15. See J. Story, 'Spain's external relations redefined: 1975–1989', in R. Gillespie *et al*. (eds), op. cit., p. 41.
16. During the Gulf war, a number of Spanish missions were sent to Maghreb capitals to allay fears about the extent of Spain's commitment to the allied engagement against Iraq. See C. Spencer, *The Maghreb in the 1990s*, London, IISS Adelphi Paper no. 274, February 1993, p. 55.
17. See A. Marquina, C. Echeverria Jesús, 'La politique d'Espagne au Maghreb', in *Monde arabe. Maghreb-Machrek*, no. 137, July-September 1992, p.47, and C. Spencer, op. cit., p. 60.
18. The Tunisian Prime Minister paid a visit to Madrid, 26–8 January 1997, within the framework of the bilateral agreement. See *Summary of World Broadcasts-BBC (The Middle East)*, 29 January 1997, p. 22.
19. K. W. Estes, 'Spain's view of Maghreb as NATO's southern flank', in *Jane's International Defence Review*, 1/96, pp. 20–1.
20. Morocco, like Tunisia, is trying to continue on a liberal path which deserves maximum support from the EU in order to stem the tide of radical Islamism in the Maghreb. Signing its partnership with the EU in April 1995, Tunisia was the first North African country to make the commitment to integrate its economy within the European sphere.
21. Algeria has the fifth largest natural gas reserves in the world.
22. Spain's catch is the fourth largest in Europe. The industry employs about 100,000 fishermen with another 700,000 related jobs. This was one of the fastest-growing sectors in the economy in the 1980s.
23. Morocco was insisting on reductions of 50 per cent to the catch, including prohibitions on fishing certain depleted species.
24. See the Spanish bilateral Declaration, issued in the afternoon of the 17 February 1989, in *Actividades, Textos y Documentos de la Política Exterior Española*, Madrid, Ministerio de Asuntos Exteriores, February 1989.
25. In 1990, the UMA countries together with France, Italy, Portugal and Spain created a western Mediterranean cooperation network, the 'Five Plus Four' grouping, which after the admission of Malta, in 1991, became the 'Five Plus Five'. The Mediterranean Forum includes Algeria, Egypt, France, Greece, Italy, Morocco, Portugal, Spain, Tunisia and Turkey,

and comprises working groups on political, cultural and economic issues.

26. F. Fernández Ordoñez, 'El Mediterráneo en busca de una estructura de seguridad', in *Revista de la OTAN*, No. 5, October 1990, pp. 7–11.
27. See *CSCE Digest*, October 1990, p. 5.
28. *Italian-Spanish Non-Paper on CSCM*, p.1.
29. Two sessions of the Inter-Parliamentary Union have taken place on the CSCM: in Málaga in June 1992, and in Valetta in November 1995. See IIe Conférence interparlamentaire sur la sécurité et la coopération en Méditerranée: *Document final*, Valett (Malta), 1–4 November 1995.
30. See C. Echeverria Jesús, 'North–South Cooperation in the Western Mediterranean', in A. Marquina (ed.), *Confidence Building and Partnership in the Western Mediterranean. Issues and Policies for the 1995 Conference*, Madrid, UNISCI Papers No. 4, 1995, pp. 9–29.
31. See 'TRAMONTANA Exercise: Mediterranean Cooperation in Action', in *Eurodefence*, no. 1, January 1995, pp. 51–2.
32. See C. Echeverria Jesús, 'La UEO y la seguridad en el Mediterraneo', in *Ejército* no. 668, December 1995, pp. 69–71, and his 'European Security and the Mediterranean: Western European Union Approach', in *Wiener Bläter zur Friedensforschung*, no. 86, 2 March 1996, pp. 56–9.
33. In September 1994 an informal meeting of NATO Defence Ministers was held in Seville.
34. See 'Don Juan Carlos reafirma el apoyo de España a la reforma de la OTAN y vincula la paz en Europa con la estabilidad mediterránea', in *España 96*, May 1996, p. 3.

# 5 The Making of Portugal's Mediterranean Policy

Fernanda Faria

Portugal's foreign policy has undergone great changes since the period of dictatorship. This could not be otherwise, given the limited and narrow scope of that period's foreign policy, which only served an isolated and isolationist regime. This chapter examines the evolution of Portugal's foreign policy from 1974 until today, taking the case of the Mediterranean, particularly the Maghreb, and attempting to understand why and how internal and external factors have influenced the unavoidable changes, as well as looking at what has changed in the way foreign policy is made in Portugal.

## FOREIGN POLICY PRIORITIES DURING THE DICTATORSHIP

From 1926 until 1974, Portugal lived under dictatorial rule. Firstly the government was in the hands of the military. After 1933 and the creation of the Estado Novo (New State), the country was ruled by Salazar, head of the government, who had far-reaching powers and control over the internal and external affairs of state. No opposition parties were allowed and only the official party were seated in the National Assembly. Thus the legislative body was deprived of its powers and was reduced to occupying a mere symbolic role in political life. All political and legislative powers were held by the government or more precisely, by the head of the government. In the meantime, Europe was undergoing major changes in terms of economic and political development, particularly after the end of the World War II.

Despite changes in Europe and the links Portugal was developing with some of the main European powers, and the United Kingdom in particular, Portuguese foreign policy priorities and orientations did not change significantly.

Foreign policy considerations were still marked by what had been the major concern of previous centuries, that of preserving national independence. Portuguese fears of annexation by its single continental neighbour, Spain, were still present, despite co-operation with the Franco regime following the Civil War in Spain. Two other major concerns for the country's foreign policy decision-makers were those of preserving the colonial empire from partition before World War II and, afterwards, protecting it from the independence movements that were quickly spreading throughout Africa. Furthermore the regime worked to consolidate its position and distance itself from political changes occurring elsewhere in Europe, such as the expansion of communist ideology in Eastern Europe, and the spread of liberal democratic governments in Western Europe.

As a consequence, Portugal's foreign policy was dominated by an isolationist tendency with respect to European political developments, which it regarded as too remote from Portuguese national interests. These interests were identified primarily and almost exclusively as those of the colonies and the Atlantic region, which was considered of fundamental strategic importance to assure Portugal's security and autonomy and as a route to the colonies. These were, in fact, the 'traditional' orientations of Portuguese foreign policy. Despite this isolationist policy, relations with the United Kingdom and Spain played an important role: the United Kingdom because it was an old ally, the main trading partner of and financial investor in Portugal, and also a major power in the sea routes to the empire. Spain, the unavoidable neighbour, was still the most feared threat: in the interest of internal political stability and integrity of the country, the Portuguese government could not afford to have a conflictual relationship with the Spanish authorities. Relations with these countries were again perceived as a matter of national interest and not in terms of engagement in European affairs. Neutrality in World War II was, to a great extent, a result of such an isolationist policy.

Portugal was a founding member of the EFTA and the Atlantic Alliance and had a trade agreement with the EEC, although this did not necessarily mean a greater commitment to Europe. As for the communist bloc, Portugal considered the USSR as a threat to its national security. Portuguese membership of NATO was perceived as an enhancement of security in the Atlantic, of which the United States had become the guarantor and the leading maritime power. This also allowed for an increased differentiation with respect to Spain, who was not a member of NATO, and for Portugal to be the single interlocutor in the Iberian Peninsula. Despite the Mutual Aid and Defence Agreement signed with the United States in 1951, Salazar feared American hegemony and US ambitions in Africa as much as he feared Spain. Portugal refused any financial assistance from the Marshall Plan for fear of internal changes and to avoid any kind of external dependency, which could sooner or later affect the regime and its colonial policy. 'Proudly alone', as stated by Salazar, probably best explains the 'state of mind' of Portuguese foreign policy at the time and the self-induced isolation of the regime.[1] Until its fall in April 1974, the international isolation of the Portuguese regime was further enhanced by the war for independence in the Portuguese colonies in Africa (which broke out in the beginning of the 1960s), the growing international pressure for decolonisation, the involvement of external powers in the war in Africa, and the growth of organized opposition forces within the country.

Relations with the Maghreb countries were almost non-existent. Not only was the Mediterranean not a priority but, furthermore, the nationalist policy of the regime tended to ignore the heritage and influence of Arab-Muslim culture on Portugal. The regime tended to reduce historic relations with the Mediterranean region to political, military and religious conflicts, over-emphasizing the period of the Reconquista (Reconquest: the fight against the Moors for control over the Iberian peninsula) when Portugal became an independent state with its current boundaries. The Maghreb governments, as supporters of African liberation movements, including those in former Portuguese colonies, received no sympathy from the authorities in Lisbon. Some Maghreb governments even welcomed opponents of the Salazar regime who were seeking political asylum during the dictatorship period.

## THE TRANSITION PERIOD: FROM 1974 TO 1986

The process of political transition that Portugal has undergone since the April 1974 Revolution which ended more than forty years of authoritarian rule has had a major impact on the internal and foreign policies of the country.

After the revolution and until 1976, Portuguese politics were dominated by the democratisation of the country's political system and by decolonisation. The period was marked by domestic political instability and a major internal crisis which came close to a shift towards a military or a radical left dictatorial regime. With the return of the country to the international fora and the re-establishment or normalisation of diplomatic relations, namely with the United States and Europe, foreign relations regained momentum in the political life of the country and were perceived by the democratic forces as of great importance for the support and consolidation of the transition process. However, it would not be until the institutionalisation of democracy and the effective functioning of representative institutions that Portugal could engage in re-building a foreign policy beyond the immediate and crucial concerns of its unstable internal situation.

From 1976 and until 1986, when Portugal became a member of the European Economic Community, Portuguese diplomacy still fundamentally concentrated on fostering political and economic support for the transition process. Though the troubled years of 1975 and 1976 were over and the democratic option was reinforced by the election of representative institutions, the main political forces in the country converged in the belief that membership of the European Community was the best guarantee for positive political evolution and a definitive consolidation of democracy in the country.[2] The main motivations for EEC membership were primarily of a political nature: Portuguese foreign policy was fully engaged in a struggle for acceptance within the European and democratic 'club' and afterwards, when membership seemed to be certain, in the economic development of the country for which international assistance, namely of the EEC and financial institutions, was sought.

Besides 'the priority of the priorities', as joining the EEC was referred to by Freitas do Amaral, the then minister of foreign affairs,[3] Africa and the search for a new framework of relations with the former African colonies was also an important trend. But it was understood and dealt with more as a matter of internal politics than one of foreign policy, given the polarisation of Portuguese public opinion regarding the decolonisation process, and particularly the civil war in Angola.

A third priority was the reinforcement of the Atlantic dimension. Atlanticism, like EC membership, was perceived as an anchor into the democratic world and in opposition to the Soviet influence in internal politics, which had been felt quite strongly in the 1974–75 period. The Atlantic dimension was developed through a closer relationship with the United States – the bilateral defence agreement with the US which had existed since 1951 was extended in 1983 – and an increased participation of Portugal in NATO, namely in its military missions. Despite the importance of the Atlantic option during the dictatorship period, relations with both the US and NATO were negatively affected by the long-lasting colonial war that Portugal was fighting in Africa. Political and diplomatic relations with the US suffered considerably then, Portuguese policy in Africa being condemned by the international society including the US. On the other hand, the colonial war also meant the non-participation of Portugal in NATO military operations for about fifteen years, its military effort being almost exclusively devoted to the war in the African colonies.

As foreign policy efforts concentrated on the three priority areas of Europe, Africa and NATO, only limited effort could be given to other areas of the world, including the Mediterranean. Despite the important improvement in relations after the April 1974 Revolution, particularly with Maghreb countries (diplomatic representations were opened or reinforced and relations were conducted at the highest political levels), it was not until the late 1980s and especially the 1990s that Portugal started developing a global policy for the region, both bilaterally and at the level of Europe's common policies.

## THE IMPACT OF THE NEW DOMESTIC AND INTERNATIONAL CONTEXT

The increasing consolidation of the democratic regime, the political stability and the economic development of the country, as well as the economic and political support given by EC membership created new conditions and new instruments for the development of a foreign policy coherent with Portugal's new political status. Furthermore, the end of the Cold War allowed and required a new approach to foreign relations, particularly towards central and eastern European countries, Russia and the Mediterranean.

Portugal had established representative institutions in 1976 that could play a role in defining, promoting and developing foreign relations, but the making of foreign policy was still essentially initiated and conducted by the government. The Parliament, particularly through its European Affairs Committee, the Committee for Foreign Affairs, Portuguese Communities and Cooperation, and the National Defence Committee, where political parties are present according to their representation in Parliament, can question and advise the government. In the case of the European Affairs Committee, it can propose or present resolution projects to the Parliament, which may ultimately become law. International treaties and conventions or defence agreements require ratification by the Parliament. It is unlikely that a foreign policy initiative requiring formal approval by the Parliament would be taken if the government knew it to be unpopular and unlikely to be ratified. Normally the government brings issues in Parliament for general discussion before it takes a final decision, to ensure that it will command the support of the majority.

The same is true regarding the role of the President in foreign policy. Though having no executive powers in foreign and defence policy (except for his right of veto), he can play an important role as the highest political representative of the country. Such a role depends very much on the personality of the President himself, rather than on the attributes of his function. Mário Soares, the former President, is an example of an interventionist and active President in external relations. The large support he enjoyed in the country and the fact that he is also well known abroad and respected for his opposition against the dictatorship, his fight for democracy and as one of the most famous politicians of the country, enabled him to play an influential role in foreign policy issues. His frequent travels abroad as well as the diplomatic initiatives he promoted, sometimes quite controversial,[4] were symptomatic of his will to participate and to play a role in the making of foreign policy. The President can thus be a lobbying force with power to influence foreign policy according to his personality, popularity and international recognition.

Public opinion also plays a role, but polls still show that, depending on the issue, a high percentage of Portuguese do not have an opinion on foreign policy and security issues; even on integration in the EEC, an issue that was the top foreign policy priority at least after 1976, polls in 1983 showed that the majority of Portuguese had no opinion about the EEC, contrasting with the large majority of the political elite who were in favour of adhesion (except the Communist party and other smaller extreme left parties). Throughout the 1980s and particularly in the 1990s, the Portuguese society has shown an increasing interest in international relations, particularly for Europe and namely central and eastern Europe, transatlantic relations, frequently for Africa, but also for Brazil, the Third World and North–South relations.[5] More direct forms of participation in foreign policy-making, such as a referendum, are still not allowed by the Constitution, but this is likely to change with a future revision of the Constitution. The debate has been focused mainly on the revision of the Maastricht Treaty, which most political parties would like to see subject to national consultation. It seems now very likely that before the end of 1997 international treaties will be subject to a referendum, now that the main political parties have reached agreement to include such a development into a revised Constitution.

Public opinion has, nevertheless, supported and encouraged, to say the least, changes in foreign policy. Along with the political transition, the Portuguese society also went through great changes: it became more open to the rest of the world, eager to catch up after the decades of isolation during the Salazar regime. The number of studies and initiatives concerning world affairs and of institutions and media interested in international political and security issues have increased, particularly since the 1980s. Public opinion became also more aware of the world's interdependence and of the likelihood of being affected by changes or events occurring in other countries or areas of the world. The scope of national interests tends to go beyond national or regional borders and become global, hence the justification for an increased participation in world affairs. The end of the Cold War has further reinforced such perceptions.

Since the beginning of the transition process there has been a large consensus on the new foreign policy priorities, particularly among the main political parties (Socialists and Social-Democrats), who tend to agree on its major guidelines. Thus, there has been little change in foreign policy priorities, despite changes of government. Opposition comes normally from the most 'conservative' parties, the Communist Party and the Popular Party (extreme right), who tend to emphasise the traditional areas of interest for Portugal as opposed to new ones in current foreign policy. They are also the ones who oppose European integration, and perceive current foreign policy, particularly in its orientations since the mid-1980s, as an alienation of Portugal's interests and identity.

European integration is another major element in the making and shaping of Portugal's new foreign policy orientations, but they have to be seen in the wider context of internal, regional and international changes.

## The 'Europeanisation' of foreign policy

The changes occurring in Portugal's foreign policy after 1986, when it became a member state, are often referred to as 'Europeanisation', as opposed to the almost exclusive 'Atlanticism' of the previous regime, in order to underline the increasing European dimension in foreign policy priorities, not only in terms of greater involvement and participation in European affairs and institutions, but also in terms of making of the main European priorities matters of national interest. The European dimension in its foreign and security policy was further enhanced by membership of the Western European Union (WEU) in 1988. Areas which were largely ignored before, such as Central and Eastern Europe, Russia and the former Soviet Republics, the Middle East and the Maghreb have gained a new importance in Portugal's foreign policy.

Membership of the EU gave Portugal an equal voice in foreign policy matters and access to substantial means and policy-making instruments which Portugal alone would not be able to mobilize or develop. The EU mobilizes much greater resources than any single European country could do, particularly Portugal given its limited financial resources and low level of diplomatic presence in many of the aforementioned regions. Besides, as a member state, the Portuguese government has the ability to take position on issues of common interest in foreign and security policy and participate in the decision-making process; it has the responsibility of doing so as part of the full exercise of its membership, of the powers and functions it implies. This sense of 'shared responsibility' is a direct consequence of EU membership and was gradually accepted and embraced by the Portuguese authorities and public opinion in general.[6] Along with it, there is a growing awareness of Portugal's new extended frontiers, now going as far as Finland and Greece, and consequently of enlarged political and security interests.

The rethinking of Portuguese foreign and security policy is also sensitive to its new condition of being an external border of the EU, a tighter control of which is required as a consequence of the loosening or, in the case of the Schengen signatories, the elimination of border controls and the growing freedom of movement in the Union. This concern has also affected Portugal's policy towards its neighbouring south, the Maghreb.

Nevertheless, the 'Europeanisation' of Portuguese foreign policy priorities did not mean just an enlargement of Portugal's foreign and security interests, nor did it imply a marginalization of the 'traditional' priorities in external relations. On the contrary, by enhancing the political role of Portugal in European and world affairs, it has strengthened its influence and prestige in relations with third countries. For instance, relations with Sub- Saharan Africa and Brazil have regained a new political importance both for them and Portugal. As a member state, the Portuguese government has powers to influence communitarian decisions. It can use these powers to best safeguard its specific interests in the Union and to promote greater European cooperation with Africa and Latin America. In fact, bilateral cooperation was reinforced and EU relations with those regions are given greater attention and priority than before, not just in terms of economic cooperation but also at the political level. The cultural and historical links with Portuguese-speaking countries in Africa and Latin America has given Portugal some comparative advantages in its attempt to play the role of bridge between these countries and the EU.

When enlargement to central and eastern European countries was first envisaged, the 'Europeanisation' of Portugal's foreign policy was also followed by a concern for a balance between interests of member states. Southern European countries feared marginalisation of their prioritarian interests in a EU increasingly continental and inward looking. Though recognising the importance of stability in its eastern frontiers and the legitimate aspirations of the new European democracies to EU membership, southern European countries, Portugal included, have expressed great concern for the risks of instability in the southern Mediterranean while pushing for the development of a common European policy for the region.

## A new strategic environment

The end of the Cold War and the collapse of the Soviet Union has brought major changes to international relations. European countries that were before in the sphere of Soviet influence have democratised and are now candidates for EU membership and Western security institutions. While current relations with Russia are shaped more by dialogue and cooperation than potential conflict, the risks of a military confrontation seem no longer credible. The threat of a global conflict has given way to regional and internal conflicts, the roots of which are in most cases of economic, social, ethnic, religious and political nature. The arms race has given way to agreements for the reduction of non-conventional arsenals, while countries are reducing their armed forces and their military budgets. Proliferation trends in some parts of the world, particularly in the Middle East and Asia, remain an important concern as the potential for regional conflicts is still very high. Besides, the end of bipolarity was perceived by some of these countries as having reduced their strategic importance which some have attempted to compensate for by an increase in military power.

Political relations with the world could no longer be guided by the logic of bipolarity, so new cooperation frameworks had to be developed. The EU has engaged in an effort to redefine its external relations, particularly with its closest neighbours to the East but also to the South, as it became aware of the negative implications instability in its immediate neighbourhood could have on its own stability and integration process. New security concerns, which are different in nature to the dominant issues in Cold War times, have emerged and are today major arguments for the new foreign policy orientations and priorities in Portugal as in the rest of Europe. Among the new priorities in Portugal's foreign relations, the Mediterranean and more specifically the Maghreb, since the Portuguese authorities consider that there are different subregions in the area, is considered a major one by every government since the mid-1980s.

In many ways, relations with the Maghreb are as much the result of the new strategic environment, as they are of integration into EU (the sharing of Europe's priorities and the availability of a set of political and economic instruments that Portugal did not have the means nor the interest to develop before in bilateral terms), not to mention of course the internal political transition that has created the conditions for the changes that Portugal's foreign and security policy has undergone.

## THE MEDITERRANEAN

The Mediterranean in Portugal's foreign and security policy means primarily and substantively the Maghreb countries. Relations with Middle Eastern countries are in general less developed and not a high priority in Portugal's Mediterranean policy. However, the region is not totally deprived of political and strategic importance for Portugal. Apart from its dependence on oil from the Gulf States which is likely to be affected by conflicts in the region, the other major sources of instability in the region are the Israeli-Palestinian conflict and the still tense relations between some Arab countries and Israel. Negative repercussions occur in the Maghreb, in the public opinions on both shores of the Mediterranean and also in relations between the West and Arab countries, especially at a time of growing political radicalism on both shores. But Portugal has never really had a policy for the Middle East (despite some diplomatic initiatives by former President Mário Soares.[7]) The diplomatic presence of Portugal in Middle Eastern and Gulf countries was very limited, which meant from the start a handicap in terms of lack of information, of knowledge and of the means to have an active policy for the region. Besides, the strategic importance of the region during the Cold War period meant that the political and strategic space was already largely occupied by the great powers. It must be said that since EU membership, Portugal's attention to and interest in the area have increased. Portugal has strongly advocated the development of a Common Foreign and Security Policy (CFSP) for the Middle East[8] and EU support of the Peace Process. Bilaterally, there are prospects for an increase in economic relations and trade, but they remain very modest, while diplomatic relations have also undergone development. Nevertheless, the Middle East is not and it is not likely to become in the near future a major priority for Portugal's foreign and security policy. Priority is still given to the Maghreb and though the distinction between sub-regions in the Mediterranean is not made by the EU, whose strategy is to take the Mediterranean as a whole, Portuguese authorities still consider that such a distinction is a valid one.

It should not be surprising that Portugal pays such great attention to the region given the geographic proximity with Maghreb countries and the historical and cultural links with the Mediterranean region and the Arab-Muslim culture, still vibrantly alive today and looked at in a much more objective and fair way than was the case under the Salazar regime. What should be surprising is Portugal's long absence from the region. However, the return of Portugal to Mediterranean politics did surprise some, who were used to thinking of Portugal as an exclusively Atlantic country, not even bathed by the Mediterranean sea. Internally there are still radical supporters of Atlanticism, who oppose Portugal's political involvement in the Mediterranean, though they are in a minority and even public opinion is now more sensitive to issues regarding the Mediterranean region and the Maghreb in particular.

The political changes in Portugal, including membership of the EU, and new perspectives on Portuguese history have paved the way for the development of relations with the Maghreb countries. Yet it is the concern for stability in the region and its implications in European and national security and stability that have provided the major impulse to developing of a global policy for the region, both at the bilateral and European level. There is a strong belief that Portugal and Europe in general are not immune to conflict or instability, particularly in the Western Mediterranean. Therefore, it is in the interest of the Union and its member states, especially those who feel more directly affected by instability in the area, to promote peace and stability.

## Threats to security and stability

Europe's main security concerns regarding the Mediterranean area are related to the economic, social and political crisis most southern Mediterranean countries face. Leaving aside the causes of their economic troubles, which are not the direct concern of this chapter, the impact on the population of the high unemployment rates, the housing problems, the deteriorating living conditions in general and the lack of prospects for a rapid improvement of the situation are all a major cause for social unrest and political instability. The corruption of the ruling elites, the non-democratic nature of the majority of the regimes in the region, the repression of minorities and non-tolerated political opposition groups on behalf of national unity are some of the other factors that have contributed to the discrediting and the loss of legitimacy of regimes and favoured the emergence of radical political movements. Some of the radical political groups do not hesitate in taking advantage of terrorist means to undermine the government and seize power, increasing the likelihood of internal conflicts and adding to mounting regional tension.

The implications of these problems for European countries are perceived first of all in terms of a higher potential for migration flows from the Southern rim into Europe, which European authorities try to control by imposing severe restrictions on immigration. The persistent economic recession in Europe, the growing unemployment rates, and consequently the diminishing capacity of European societies to absorb extra labour force, the lack of integration policies, and the more frequent demonstrations of xenophobia and racism, have all made the issue of immigration an EU policy priority in the 1990s.

Though Portugal is not a common destination for Maghrebi migrants (the main immigrant communities are from former African colonies), it fears that continuous migration flows will end up affecting Portugal too and it has adopted the most restrictive measures being applied within the EU. Besides, as an external frontier of the Union, Portugal has additional responsibilities in the control of its borders with regard to its European partners. Such responsibilities, probably more than any immediate and direct concerns regarding Maghrebi immigration, have contributed to Portugal's commitment to the Schengen agreements and the influence of Charles Pasqua's policy when interior minister of France, which tended to perceive in each immigrant a potential terrorist threat. Portugal also participates in the initiatives promoted by the French government, also initiated by Pasqua, for coordination with other interior ministries in Southern Europe in the fight against international terrorism, arms and drug trafficking, activities that often seem to be connected.

Political instability and radicalism in the Maghreb countries also have an impact on its migrant populations in Europe, as well as on European public opinion. Again, those countries with a large presence of Maghrebi immigrants in their territories are likely to be more affected than other member states – the case of France and Algerian relations being an example of contemporary tensions. However, the impact on European public opinions of the political radicalism and growing violence in the region, the authoritarian nature of regimes, and numerous violations of human rights, put European governments under pressure to develop a policy for the region that may promote and reinforce the respect for human rights and democratic principles, which are perceived as the only guarantees of long-term stability. The dilemma Portuguese and southern European governments face, certainly more than other member states, is how to promote political reforms in the absence of strong democratic institutions without aggravating political tensions or further undermining the current pro-western regimes, making them even more vulnerable to the radical Islamist opposition movements. The strategy of southern European governments, and now that which the EU advocates, is a gradual transition process that may allow the consolidation of economic reforms, while trying to minimise their social impact, and the development of more favourable conditions for a peaceful and positive political transition.

The Portuguese authorities are particularly sensitive to the social and political problems of any transition process, as they have recently experienced one themselves, and the current socialist government is probably the more so, also for ideological reasons. It is thus not surprising that the emphasis of southern European governments is very much focused on the development of EU economic cooperation, integrated in a more global and coherent policy for the region that may take into consideration the social, political and security aspects.

Though security and defence concerns are not a major priority in European policies for the Maghreb, following the end of the Cold War, they have been subject to more attentive analysis. Various political initiatives such as mutual perceptions of threat have dominated some political rhetoric on both shores of the Mediterranean. The perceived temptation of the West to create a substitution enemy for the no longer existent communist threat (the Gulf war is perceived by Arab public opinion as a sign of such a trend) and the rise of radical, anti-western, Islamist movements in the southern Mediterranean have largely contributed to create such perceptions of threat, which are not, or at least no longer, the dominant ones. There is currently a better and more comprehensive understanding of problems in the region.

Though it is widely recognized in Europe that there are no military threats directed from the region, there are however a number of factors that represent potential threats to security and are likely to have defence and military implications. Security problems in the region are primarily of South–South nature. They were firstly related to border disputes, which were the origin of military conflicts and tensions between Maghreb countries, not yet completely surmounted, as the Western Sahara conflict and the enduring tension between Morocco and Algeria, or Tunisia and Libya, clearly demonstrate. More recently, security problems in the Maghreb are predominantly related to internal problems that threaten the domestic stability and unity of some Maghreb countries and may ultimately lead to their disintegration and trigger new conflicts in the region: for instance, the case of ethnic and religious conflicts in Mauritania or the confrontation between radical Islamists and the army in Algeria which has led to a quasi-civil war. In other countries, the potential for internal conflicts cannot be ignored either, as social and political tension arises. Radical Islamist movements are playing a major part in this new context, not only their role as internal destabilisers, but also as factors of tension and potential conflict between countries in the area.[9]

In the past, some of these conflicts have affected European and specifically Portuguese interests, as was the case during the Western Saharan conflict when Portuguese fishing boats were attacked by forces of the Polisario Front. Today, conflicts in the area would have a greater effect, given the significant economic and financial interests European countries have in the region, as well as the important human and political links between both shores. The rise of radical Islamic movements and the possibility of an hostile, anti-Western regime are of special concern to Europe, for their impact on migration flows, as well as on migrant communities in Europe. This could become an internal problem for some member states, as events in France since 1995 have illustrated. Moreover, signs of recruitment and arms trafficking on European territory by Islamic movements and the possible expansion of terrorist acts, as was the case in France, are of great concern to European governments, Portugal included.

Another concern relates to the proliferation of weapons of mass destruction and of ballistic missiles. This is particularly important in respect to Middle East countries. In the Maghreb, it is less likely that proliferation might be a major issue, but there is still the case of Libya and much depends on the prospects for long-term peace and stability in the area. However, there are attempts to address these and other hard-security and defence-related issues through political initiatives and dialogue with Maghreb and other southern Mediterranean countries. Portugal has been one of the promoters of some of these multilateral initiatives with some southern Mediterranean countries, such as the WEU initiative in 1993.

**Bilateral policy**

Though the economic dimension of Euro-Maghreb relations is probably the most developed of European policies for the region and still considered by Portugal and other member states as insufficient to cope with the economic and social problems, it is not Portugal's most developed policy in the Maghreb. On the one hand, since the eighteenth century, since the loss of the last possession in the Moroccan coast, Mazagan, Portugal's interests and influence in the region have been insignificant. It was only in the 1980s that Portugal started developing a policy for the area. Until then, diplomatic representations of Portugal in southern Mediterranean countries and vice-versa were either non- existent or reduced to a minimum. On the other hand, Portugal is more limited financially and in resources to develop a strong policy of investments in the region than are other southern European countries.

Instead, the Portuguese government tries to profit from what it considers are Portugal's political comparative advantages in relation to its European partners, in particular at times of more intense but also troubled relations between the two shores: it has no territorial disputes (as Spain has, because of Ceuta and Melilla), no recent colonial past (unlike France, Italy and Spain), and no major economic or financial interests or strong human links that may bias its political attitudes or be looked at with mistrust.

The Portuguese authorities have thus invested primarily in political relations, while trying to develop economic relations too. Though Portugal's interest in the region started prior to European integration, membership has made available means and diplomatic instruments Portugal alone could not afford or develop. Membership also reinforced its political influence, especially with regard to Maghreb states, to whom Portugal became another 'lobbying force' in favour of a more developed common Mediterranean policy, once the first clashes of interest regarding some agricultural products and particularly textiles were overcome. Diplomatic representations have been established or reinforced; official visits have become more frequent and in some cases annual summits between chiefs of state have even been institutionalized; bilateral cooperation agreements or protocols (on cultural exchange, education and others social and political areas) have been signed between Portuguese and Maghrebi institutions; defence cooperation agreements (on technical assistance and the training of officers) were signed with Morocco and Tunisia in 1993 and 1995, respectively. At the same time, Portugal takes part, as a founding member, in various multilateral initiatives for the Mediterranean created during the 1990s .

However, in Portugal's bilateral relations the same priority is not given to all Maghreb countries. Morocco has always been considered a case for a special relationship. There is only 190km of distance between the two countries (less than the distance between continental Portugal and the islands of Madeira and Azores). Morocco is Portugal's closest neighbour in the Maghreb and Lisbon the closest European capital to Rabat, as both countries' authorities often like to recall. Historically, Morocco is also Portugal's main connection to the Mediterranean. During the expansion period in the fifteenth century and until the eighteenth, Portugal had various possessions along the Moroccan coasts; strategically, Morocco was then more important to Portugal than any other country on the southern rim. Firstly because of its location at the entry to the Mediterranean and secondly in case of an attack from the continent, Morocco was seen as a basis of support for Portugal's defence system. Which explains the interest of Portugal in blocking the Ottoman empire from occupying what is today Moroccan territory, besides the religious beliefs or political interests that were at stake. The strategic environment is a different one today, but stability and the existence of a friendly and pro-Western government in the closest Maghreb neighbour is still considered of great importance.

Despite the geographic proximity and the political priority given to the Maghreb, economic relations remain at a very low level. Trade is insignificant representing only 5 per cent of Portugal's foreign trade; exports are less than half of the exports,[10] of which 80 per cent are of oil and related products imported from Algeria and from Libya, who is ironically an important economic partner in the region, though political and diplomatic relations are almost non-existent. Instead, relations with Morocco are less significant, even if Morocco is the main destination of Portuguese exports to the region (about 40 per cent). Portuguese investments in the Maghreb and vice versa are also very low or non-existent.

However and thanks to the energy sector, economic interests in the region are growing quickly, thus reinforcing the political links between Portugal and the region. The Portuguese government has made large investments in the construction of the gas pipeline Maghreb-Europe (and related infrastructures) that will bring natural gas from Algeria to Spain and Portugal, increasing Portugal's dependency on energy imports from the Maghreb, which are expected to meet 10 per cent of Portugal's energy needs by the year 2000. This already has an important impact on the making of foreign policy towards the Maghreb and the concerns of Portuguese authorities for stability in the area.

Trade in general is also growing. From 1991 to 1995, exports grew by 250 per cent while imports augmented 100 per cent and will certainly increase when the gas pipeline comes into operation. Equally, Portuguese investments in the region are also expected to grow; in 1992–93 the government created credit lines to promote private investments and trade relations with countries in the region, so much below the level of what would be politically desirable. There is also a change in the attitude of the Portuguese authorities towards economic cooperation. They tend to explore the similarities of national and Maghreb economies, considered to have a low degree of complementarity (both in textile industry, fisheries and agricultural products), and develop co-operation at the levels of industry, infrastructures and technical assistance. Portugal's recent experience of economic reforms and integration in the European market is also considered a valuable element for the promotion of the economic as much as the political relations with the Maghreb countries, which are preparing to open up their economies and national markets to European industrial goods. From time to time, Portuguese authorities still raise obstacles to the economic agreements of the EU with non-member Mediterranean countries, as it happened at the end of 1995 when Jaime Gama, the current minister for foreign affairs, briefly blocked the negotiations of the association agreement with Morocco for a matter of fisheries rights. However, such obstacles are more the result of deep crisis in traditional sectors of Portuguese economy which need large investments for restructuring, conversion or modernization, than an opposition to the opening up of the European market to southern Mediterranean goods.

**From multilateral initiatives to a common European policy**

Since the beginning of the 1990s Portugal has participated in various multilateral initiatives proposed or co-sponsored by southern European countries. The purpose of these is to promote cooperation with Maghreb states at the economic, socio-cultural, political and security levels, and to act as a lobbying force within the EU to promote the development of an integrated common policy for the Mediterranean and not just a limited trade-oriented policy. Since its Presidency of the European Council, in the first half of 1992, Portugal has been at the forefront of political initiatives for the Mediterranean within the EU and other European security institutions.

The first of these initiatives was a proposal for a project and a Conference on Security and Cooperation in the Mediterranean (CSCM), for the whole Mediterranean region involving Russia and the United States and comprising a security and defence dimension. But it was not implemented given unfavourable political conditions and the obstacle represented by the Israeli–Palestinian conflict.

Instead, the 'Five plus Five Group' was created in October 1990, and it was limited to the five states of the Arab Maghreb Union (Mauritania, Morocco, Algeria, Tunisia and Libya) and five southern European countries (Portugal, Spain, France, Italy and Malta, who joined in late 1991). Though, similar in its main purposes to the CSCM, the 'Five plus Five Group' gave more emphasis to economic, political, social and cultural cooperation in the Western Mediterranean than to security and defence aspects, thought to be less of a priority in relations with the Maghreb and in any case different from the security problems in the Middle East. It reflected nevertheless the security interests of the southern European countries: first of all, the impact of instability and conflict in the Maghreb region on European stability and, secondly, some EU member states have defence agreements and have given security guarantees to some Maghreb states.

A third initiative, the Mediterranean Forum, was born after the 'Five plus Five Group' was paralysed following the embargo on Libya, accused of being responsible for the terrorist bombs that caused the crash at Lockerbie and that of the UTA aeroplane, and after the outbreak of the Algerian crisis. Its members are four Maghreb states (Libya is not included), Egypt (who promoted the initiative), and the southern European countries, including Malta, Greece and Turkey.

For the Portuguese government, the 'Five plus Five' framework reflected more the specific relations between Europe and the Maghreb, quite different from those with the Middle East, where the United States is perceived as the main power. Still today and despite the support given to initiatives focusing on the whole Mediterranean area, the Portuguese authorities privilege the restricted regional approach and see no obstacle for its co-existence with enlarged approaches, CSCM style, or independently of a EU global policy for the region in the way that the follow- up to the Barcelona Conference is expected to develop.

The main significance of these multilateral initiatives is political, though on their agenda there are some very concrete cooperation projects, for the most part not yet materialised. These initiatives were the first steps towards a multilateral dialogue and cooperation between European and southern Mediterranean countries; they served as a forum for debate and political dialogue and understanding; they strengthened the sense of common interest among the southern European countries and contributed to creating a lobbying force within the EU. They ultimately led to the Association Agreements with southern Mediterranean states and to the Euro-Mediterranean Conference in Barcelona. These were the EU responses to pressure by southern European countries for an enhanced Mediterranean policy not only limited to economic terms but including political and security dimensions, both necessary to cope with the growing challenges in the region. It remains to be seen what common foreign and security policy will come out of the follow-up of the Barcelona Conference, assuming that the institutional reform of the Union in discussion at the Inter-Governmental Conference will create the minimum conditions necessary for the development of a CFSP.

Unlike its bilateral policy for the region, Portugal's efforts for the development of EU Mediterranean policy are much more focused on economic aspects, and for obvious reasons. Upgrading the multilateral initiatives to a level of a common European policy meant for southern European countries the possibility of mobilising more financial means and political instruments, besides their enhanced political credibility and the strengthening of southern European countries political roles within the Union, as opposed to the marginalisation they feared so much as the EU was looking eastwards. At the Barcelona Conference, Portuguese proposals promoted the liberalisation of investment and the creation of new financial instruments in order to facilitate economic cooperation. The hope was that such measures will help the industrial development and the creation of jobs in the region. However, Portugal has not been less active in the promotion of the political and security dimension of Euro-Mediterranean relations.

It was during the Portuguese Presidency of the EU Council in 1992 that the Maghreb and the Middle East were considered areas for common action. The security dialogue that the WEU initiated in 1993, firstly with Maghreb countries and then extended to other southern Mediterranean countries, on an individual basis, was originally proposed by Portugal. In 1995, NATO started a similar dialogue, following a Spanish proposal, supported also by Portugal, despite the initial caution for fears of negative reactions from the southern Mediterranean countries. More recently, the Portuguese government has been one of the strongest supporters of a closer relationship between NATO and the southern Mediterranean states. It has been advocating particularly the interests of Morocco in such a relationship. To quote the minister for foreign affairs, 'there is plenty of room for the future development of NATO's Mediterranean initiative . . . into a more fruitful partnership encompassing joint exercises and training. And the time may come when PfP [Partnership for Peace] or a PfP inspired model will be extended to this region'. He adds that given its proximity, the 'Iberlant [Nato sub-command nearby Lisbon] finds itself in an excellent position to play a part in this cooperation, which could prove to be an element of stabilization of the Mediterranean's southern rim.'[11]

Besides concerns for stability in the Mediterranean, fears of marginalisation of southern European countries, as Europe was looking to the East after the disintegration of the Soviet bloc and the fall of the Berlin Wall, have largely contributed to the growing number of initiatives for the Mediterranean region promoted by southern European states and the understandable pressure on all EU member states not to ignore the southern neighbours on the other side of the rim. However, this urgent need to come up with political formulas has led to a multitude of initiatives, often looking like a remake of the European policies for central and eastern Europe, though Europe is well aware of the different reality of the southern Mediterranean states. From the CSCM to Malta's proposal of a Stability Pact for the Mediterranean or the allusion of Portugal's foreign minister to a Partnership for Peace or an inspired model for the region, EU policy to the East is still the main inspiration for southern European countries and what gives most drive to the illusion of equal priorities and policies towards the East and the South. However, their goals are not exactly the same – even if stability and development is a fundamental goal, which is common to policies for both regions – and they may generate expectations in southern Mediterranean countries that member states will not be able or willing to fulfill or to give priority. These also include matters that non-member Mediterranean states themselves may not be prepared to address or engage in. Notwithstanding the declared awareness of the differences between the East and the South, there seems to be little effort to clearly specify the contents of such cooperation initiatives for the Mediterranean.

## CONCLUSION

A few final remarks are required, both regarding the evolution in the content and the making of Portugal's foreign policy from pre-revolutionary times until today.

First, within the process of change there are elements of continuity, that we often refer to as elements of tradition. In terms of foreign policy priorities, these are the European integration process, relations with former African colonies and trans-Atlantic relations. Though looked at in a quite different way from what they represented during the dictatorship period, they are nevertheless perceived as a reference to traditions in Portugal's foreign relations and identified with its most immediate national interests and even, in some political rhetoric, deeply related to its national identity. In relation to the new areas of interest in Portugal's foreign policy, these priorities are still the first. However, with the increasing complexity of the European integration process and the evolution towards a political union with a common foreign and security policy, it will be more and more difficult to separate the main priority of European integration from areas of immediate concern and interest of the Union. The European integration priority may end up representing a number of areas and issues that before were considered separately or simply not part of the country's main foreign policy interests. The case of the Maghreb and the Mediterranean in general is symptomatic of these changes and of the interactions between traditional areas and emerging priorities. The same is likely to happen in respect to trans- Atlantic relations with an eventual restructuring of NATO which is needed in order to cope with a new strategic environment. For instance, the participation of Portuguese military forces in the IFOR with a contingent as numerous as the ones participating in the peace process in Angola and in Mozambique can be considered quite significant: not so long ago, that would have been unthinkable or at least very unlikely to happen. A clear definition of the priorities within priorities is thus likely to become more difficult and complex.

Secondly, the change from a one-man regime to a pluralistic and democratic political system has created or allowed new forms of intervention and participation in the making of foreign policy. However, in Portugal, the making of foreign policy still remains very much a governmental affair, in which the Parliament, the President, public opinion or interest groups play a very limited role. The Minister of Foreign Affairs is normally the leading and most influential personality in the making of foreign policy, his previous political experience being often an important reference to the political orientations he is likely to follow or privilege. For example, the current Minister, Jaime Gama, is among the first to have given greater importance to relations with Maghreb countries and supported the development of a policy for the region when he was a member of the government in the early 1980s. The priority he gives today to the region is thus not surprising as he seeks to reinforce what previous governments had developed. It is equally not surprising that he further emphasizes relations with Morocco by enhancing its role of a 'privileged' neighbour and interlocutor in the Maghreb region.

In relation to the foreign policy priorities of previous governments, there are no significant differences of political orientation, nor do there seem to be differences in terms of the policy making culture. The difference between the current and the previous government in foreign policy is generated by the more affirmative stance of the Socialists in European politics and in the defence of national interests. As strong supporters of the development of the Union into a political entity with a foreign and security policy, it is a major concern of the current executive to be in the first group of the integration process, perceived as a determinant factor to control and influence in a more decisive way the future of national and European politics.

## NOTES AND REFERENCES

1. On the foreign policy of the Salazar regime, see F. Rosas, 'O Estado-Novo (1926–1974)' in J. Mattoso (ed.), *História de Portugal*, vol.7, Círculo de Leitores, Portugal, 1994, pp. 295–9; see also J. Marques de Almeida, *Portuguese Security Policy: Between Geopolitical Culture and Institutional Commitment*, Lumiar

Papers, no. 4, IEEI, Lisbon, 1995, pp. 12–20, which refers to the 'geopolitical approach' dominant in Portugal's foreign and security policy until the 1970s under by Franco Nogueira, minister of foreign affairs during the rule of Salazar and probably the most famous geopolitical thinker of the regime.

2. Cf. Alvaro Vasconcelos, 'Portugal and the European Political Cooperation', *International Spectator,* vol. 26, no. 2, IAI, Rome, 1991; A. L. Sousa Franco, 'O Tempo Critico (1976–1985)', in António Reis (Coord.), *Portugal, 20 anos de democracia,* Circulo de Leitores, Portugal, 1994, pp. 206–57.
3. Quoted in A. L. Sousa Franco, op.cit., p. 255.
4. As when, during the Gulf War, the President sent his 'chef de cabinet' to Tunis to deliver to Yasser Arafat in person his reply to a letter he had sent him asking for European mediation, which caused a minor quarrel with the government and a protest from Israel.
5. Cf. José Medeiros Ferreira, '*Portugal em Transe (1974–1985)*' vol. 8 of, *História de Portugal*, op. cit., pp. 149 and 173.
6. An example that can be taken to illustrate the growing feeling of 'European responsibility' and engagement of Portugal in non-traditional areas of national interests and foreign policy is its participation in the western coalition during the Gulf War. For an analysis of the reactions of Portuguese authorities, political parties and public opinion, see Alvaro Vasconcelos, 'Portugal, the Gulf Crisis and WEU', in N. Gnesotto and J. Roper (eds), *Western Europe and the Gulf*, WEU Institute for Security Studies, Paris, 1992, pp. 109–25.
7. See note 4.
8. See the conclusions of the Portuguese Presidency of the European Council, in Lisbon, June 1992.
9. Islamist movements are often perceived as instruments manipulated by unfriendly neighbouring countries whose purpose is to undermine their internal stability. The Tunisian government, for example, accused Colonel Gaddaffi of Libya of training and supporting radical Islamists of Tunisia and Algeria and of fomenting instability in the country. Iran and Sudan have also been accused by several countries in the Arab world of supporting Islamic movements in the Middle East and the Maghreb. On the other hand, there are also fears that Islamic movements try to export their terrorist activities into neighbouring countries.
10. In 1994, Portuguese exports to the five Maghreb countries were worth less than 30 bn escudos (approximately $195m) and imports 70 bn ($460m).

11. J. Gama, 'Portugal and the transformed NATO' in *NATO Review*, no. 4, July 1996, p. 6.

Other useful sources are:

F. Faria, 'The Mediterranean: a new priority in Portuguese foreign policy' in *Mediterranean Politics*, vol.1, no. 2 1996, pp. 21–230, J. Marques de Almeida, *Portuguese Security Policy: Between Geopolitical Culture and Institutional Commitment*, Lumiar Papers, no. 1, IEEI, Lisbon, 1995, A. Vasconcelos, 'Portugal: A case for an Open Europe' in F. Algieri and E. Regelsberger (eds), *Synergy at Work. Spain and Portugal in European Foreign Policy*, IEP, Europa Union, Bonn, 1996, pp. 111–36.

# 6 The Model of Foreign Policy-Making in Greece: Personalities versus Institutions

P. C. Ioakimidis

## INTRODUCTION

Greece's foreign policy choices and objectives have, at least until very recently, been a puzzle to many Europeans and other international actors. Greek foreign policy pursuits have been described as 'irrational', 'parochial', 'incomprehensible', 'aggressive', even 'crazy'.[1] No other member state of the European Union (EU) has in the recent years elicited such harsh criticism as Greece has, largely because of its foreign policy objectives, strategies and actions. But even within Greece itself, the foreign policy pursued since the end of the Second World War, has been described by no less a person than a former Prime Minister, Mitsotakis, as 'a series of mistakes and disappointments' and totally counter-productive to promoting Greece's national interest.[2] Even Greece's current Prime Minister, Simitis, has condemned Greece's foreign policy as 'catastrophic to Greece's genuine interests'.[3] Likewise, the academic and professional opinion is almost unanimously criticizing Greece's foreign policy as 'ineffective', 'erratic' or 'spasmodic'.[4]

This is a curious phenomenon in that policy-makers and objective, independent analysts alike concur in condemning the foreign policy outputs and the effectiveness of the policy which these policy-makers have actually produced. In no other European country has a phenomenon like that been observed. The question which naturally arises is, Why? Why do foreign policy objectives, choices and outputs meet with such severe criticism and even condemnation even by those who have contributed to bringing them about?

To answer this crucial question, one has to start with a fundamental statement/hypothesis; namely that foreign policy-shaping in Greece hardly conforms to any known models of making foreign policy. It is an idiosyncratic model peculiar to Greece's political, cultural and historical environment. It is nonetheless a model which provides very interesting and useful parameters to enrich the theoretical work concerning the study of foreign policy-making process and analysis.

Established models of foreign policy analysis fail, in our view, to accord the proper weight to three very important factors influencing, if not determining, foreign policy outputs: first,the society in its broader shape and public opinion in particular as a source of foreign policy; second, the role individual personalities (i.e. ministers of foreign affairs) might play in shaping policies and making macro-political choices; thirdly, the interplay between personalities and society/public opinion as a factor influencing foreign policy decisions and choices.

It is characteristic that essays by two distinguished theorists on foreign policy analysis, Hill[5] and Clarke,[6] make no reference at all to the role of personalities in the process of shaping foreign policy. It is as if policy choices are made either solely by impersonal institutions or by elites of various kinds, either political or bureaucratic. In this model, the impact the foreign minister can have on foreign policy outputs is almost completely ignored or submerged under the 'elite factor'. As we will try to show below, this is plainly misplaced in the Greek case. As far as public opinion is concerned, the view expressed by Hill is quite indicative:

> Almost all the research directed towards public opinion has concluded that it is of peripheral importance in the area of external affairs the general public is seen as having little or no interest in foreign affairs – elections usually being fought on tangible, domestic issues – no expertise, and relatively little opportunity to influence policy, because of the norms of secrecy and raison d'état which prevail.[7]

As we will argue in this chapter, Greek foreign policy-making and Greek foreign policy choices can be properly accounted for only by seriously taking into consideration three factors: 'public opinion', 'the role of personality' and the 'interplay' between personalities and society/public opinion. Public opinion is of course a very transient and elusive phenomenon constantly changing and constantly taking different mutations depending on conjunctural factors and influences. Yet public opinion reflects deeply-ingrained cultural features, attitudes, historical legacies, myths and traditions. In this sense, political culture can be a powerful source of foreign policy and not merely 'affect the socialization of decision makers' as Clarke asserts.[8] It is quite possible for decision-makers to have been socialized in a certain cultural environment and yet pursue policy objectives which reflect cultural orientations of an entirely different kind.

As already noted, in Greece the making of the foreign policy process seems to correspond to a model which puts at the centre of it three distinctive parameters:

- the role of personality – the minister of foreign affairs (or the prime minister) as foreign policy-maker as opposed to that of collective bodies, institutions,
- the society/public opinion as a source of foreign policy,
- the interplay between society and personalities in the foreign policy domain.

The predominant factors in this model are on the one hand the personality and on the other the society. In this model, therefore the role of institutions and bureaucratic structures as factors in defining the policy-making process appears limited or in some cases even non-existent. Indeed, the fundamental observation from which our analysis starts, and upon which our thesis is constructed, is the absence of any effective institutional structures, procedures and processes – systemic institutional framework – for public policy-making and the consequent absence of effective collective bodies of foreign policy-making. This 'absence' does not refer of course to formal constitutionally or legally defined structures for making policy. As a matter of fact, the formal legal and constitutional arrangements provide for an elaborate institutional system of policy-making of a collective nature, but just as in several other policy areas, the formal, typical structures indicate very little about the actual process of policy-making. In the area of foreign policy in particular the gap between formal structures and the actual reality is a huge one.

Starting from this observation, the process of foreign policy-making in Greece seems to conform to a model which involves, somehow schematically, a sequence of the following stages:

(1) The absence of effective, legitimised institutional structures (systemic institutional framework) for making policy leads to the dominance of personalities in the process of making foreign policy.
(2) This, in turn, leads to making foreign policy using as paramount criterion the maximisation of the electoral appeal of the personality concerned, something which makes the society into a prime source of foreign policy.
(3) This results in making maximalistic, largely irrational foreign policy choices which very often give rise to a 'crisis situation'.
(4) The maximalistic/irrational objectives bring Greece's foreign policy into conflict with (i) the European Union (EU), (ii) the international community at large, (iii) the most advanced economic interests of the country, which adhere to rational economic behaviour and even (iv) members of the governing elite.
(5) As a result, Greece becomes thoroughly isolated in the European/international system.
(6) After a period of intransigence, during which these conflicts become exacerbated, it follows a quest for the inevitable compromise.
(7) The compromise is usually at a lower threshold than otherwise might have been the case.
(8) The compromise is liable to be received by the society as a whole, as a humiliating retreat in the face of external pressures. This tends to reinforce a definite cultural predilection thereby perpetuating a vicious circle.

These eight stages are summed up diagramatically below and are then analysed in turn.

### *THE GREEK FOREIGN POLICY MODEL*

(1) *Lack of systemic institutional framework → leads to*↓

(2) *Dominance of personalities in policy-making → leads to*↓

(3) *Maximalist/irrational decisions → leads to*↓

(4) *Foreign policy in conflict with*

The European Union (EU)
The international community
The advanced economic interests
Members of governing party
→ leads to↓

(5) *Isolation of Greece → leads to*↓

(6) *Search for compromise → leads to*↓

(7) *Compromise at a lower threshold → leads to*↓

(8) *Reinforcing traits making institutionalisation difficult*

## THE LACK OF INSTITUTIONAL STRUCTURES

The lack of consolidated institutional structures, or systemic institutional framework for shaping policy and making decisions, represents the single most important problem in the area of public policy-making in Greece.[9] As noted, the absence of such structures does not relate to a constitutional framework or to legally established organs for making policy. These organs are formally in existence. The point is that the level of institutionalisation in practical, operational terms is very weak and ineffective. The formal institutional system and the actual policy-making process have very little in common. The formal system provides for specialised cabinet bodies to make the broad foreign policy choices and decide long-term strategies.

Nevertheless research and personal accounts from former prime ministers reveal that rarely, if ever, have these collective institutional bodies actually made concrete policy choices. Indeed the institutional bodies of collective nature only meet very rarely and when they do so, they tend to engage in declaratory, rhetorical type discources, which have little bearing on actual foreign policy outputs. Especially under Papandreou's governments, in the 1980s as well as in the early 1990s (1993–95), the collective bodies of policy-making used to perform a legitimising role of policy decisions taken elsewhere.[10]

The Ministry of Foreign Affairs (MFA) is by far the best organised department of the Greek administrative system. But, like the rest of the administration, the MFA is also victim of the general pathology affecting the administrative machinery.[11] While its organisational structure corresponds more or less to organisational patterns encountered elsewhere in European Ministries of Foreign Affairs, the internal dynamics, interpersonal relations and policy roles and profiles are shaped by political considerations of clientelistic nature. As a result, the internal organisational structure of the MFA does not indicate the actual role that the bureaucracy plays in the policy-making process.

The diplomatic service views its role as that of implementing policy choices made by the political leadership of the Ministry rather than being a source of alternative policy options.[12] Thus the diplomatic staff rarely submits alternative policy scenarios of a strategic nature to the political heads of the Ministry and certainly never questions the strategic policy options chosen by the political class.[13] In the rare cases that the diplomats objected to the chosen policy course, they quited the diplomatic service altogether in order to express their antithesis.[14]

The prevailing view is that the role of the diplomatic service is not to act as a source of policy initiatives or policy outputs, but rather as an executive instrument to carry out and support with technical argumentation the policy options of the political leadership of the Ministry. As a result, 'proceduralism' is the phenomenon which predominates in the Greek Ministry of Foreign Affairs. In meetings involving the diplomatic service the contribution of the diplomatic staff revolves normally around the 'procedures', or 'procedural steps' required to maximise certain goals. In fact, this is a deeply ingrained characteristic of any diplomatic service, but in the Greek case it assumes disproportionate significance precisely because of the aversion the diplomatic service feels towards becoming involved in the process of defining policy options. The typical reaction, when confronted with such an eventuality is that it is a 'political question' which should be resolved by the political leadership and not by the bureaucratic personnel.

It is quite interesting to note in this connection that no policy planning unit has ever functioned properly in the Greek Ministry of Foreign Affairs. A unit with that aim, which had been established in 1990, was soon deactivated because neither the political leadership paid any attention to its suggestions, nor the diplomatic service wanted to become actively involved in its work. Indeed, the unit was composed of academics and retired diplomats with the exception of its head who was a senior member of the diplomatic service towards the end of his career. The specialised departments in the MFA resented (and resisted) the establishment of this unit for, as one diplomat put it, 'it risked politicising the diplomatic service by drawing it into the process of making choices or identifying alternative policy scenarios, a task that properly belongs to the domain of political leaders'.[15] Notwithstanding professional self-interest, the diplomats tend to oppose the establishment of the policy planning department because the latter's function is sharply at odds with the ideological profile they wish to project for the service, namely as a non-political actor implementing or supporting political decisions.

Nevertheless this ostensibly non-political stance results in actual fact in a highly politicised diplomatic service, albeit in a disguised form. This is because, as the diplomatic echelons appear eager to support unquestionably whatever policy options are initiated by the political leaders of the Ministry, they come to be seen as fully identified with individual political personalities.The phenomenon of politicisation is reinforced by the propensity of the political parties to establish 'domains of political influence' within the diplomatic service, thereby creating clientelistic type networks within the diplomatic service with individual diplomats linked to parties or political personalities.

Seen against this background, it is only inevitable that promotions, transfers, postings abroad and general career advancement come to depend largely on the political links established and to be decided on the basis of political criteria. It is common knowledge in the MFA that the so-called 'personnel review boards', which are supposed to take decisions on professional matters (promotions, transfers, etc.) serve only to sanction decisions made by the Minister's office or the Minister personally, especially with respect to personnel at senior ranks.

Given these circumstances, it is not altogether surprising that the diplomatic service is reluctant to undertake the function of presenting alternative policy scenarios and options and prefers to mould the image of a narrow bureaucratic body, competent to implement and support policy decisions taken by the political leadership.

What this brief analysis seeks to indicate is the absence of a systemic institutional framework for formulating foreign policy. Neither at the bureaucratic level nor at the political (governmental) level are there collective bodies which can be said to form a structured institutional system for producing policy. The net result of this is that key desicions are made by individual personalities and, as a result, they reflect personal preferences.

No doubt, the 'institutional deficit' at the political level reflects deep-seated societal features that merit a brief mention. To begin with, one powerful element that distinguishes the Greek society from other European societies is the all-pervasive sense of 'a society being threatened by outside forces' either of the immediate vicinity or even from more remote powers, including the superpowers of the day. This deeply ingrained perception of threat to Greece's territorial integrity and independent existence emanates as much from geography as from the traumatic historical experiences relating to Greece's long and painful process of establishing its national integration. It is of course a perception greatly compounded by the chronic instability prevailing in the region, as well as by Turkey's policy of questioning Greece's sovereignty, especially in the Aegean, and the notion that outside powers tend to side or openly support Turkey in this.[16]

The powerful perception of threat coupled with the societal and political consequences left by the country's long and erratic process of state formation have generated those psychological and cultural traits that render immensely difficult the moulding of stable, workable, impersonal institutional structures of policy-making, power-sharing and allocation. Mistrust and suspicion towards the international system at large are the most salient characteristics that shape state–society relationships forstering a strong mentality of subservience to state authority or to any form of authority.

These characteristics constitute an integral part of what Diamantouros has called the 'underdog culture', which has for very long predominated in Greece.[17] This culture comprises, among other features, a pronounced introvertedness, a powerful statist orientation, 'a preference for small and familiar structures compatible with the unmediated exercise of power and closely associated with the clientelistic practices'. Moreover, a pronounced element of xenophobia forms a strong dimension of this culture, an element involving a preference for conspiratorial interpretation of events, an insecure sense of nationalism, a siege mentality. It is, as Mouzelis argues, a 'political culture dominated by particularistic, personalistic discourses concealing or rendering peripheral substansive issues of social reform'.[18]

At the heart of this culture one can discern the pervasive lack of trust in interpersonal, societal relations. Recent changes notwithstanding, trust in Greek society hardly extends beyond the scope of the closely-knit structure of the family and of the intimate group of very close friends forming what is known in Greece as the 'parea' (personal company). Trust is therefore confined to a very small group of persons linked either by family ties or by the friendship bonds. Mistrust is what prevails, trust is what is lacking.[19]

Trust, defined as 'the expectation ... of regular, honest and cooperative behaviour based on commonly shared norms' is an indispensable psychological prerequisite for large scale organisations, be they economic or political.[20] Fukuyama has recently shown that low trust societies rely heavily in their economic organisation on family type, fragile enterprises or public enterprises, while in high-trust societies long-lived professionally managed large-scale enterprises predominate. Similarly in high-trust societies a thick layer of intermediate organisations (voluntary associations, educational institutions, unions, etc.) stand between the individual and the state. In low-trust societies by contrast the state tends to atomise the society and thus it renders the latter more amenable to manipulation and mobilisation. In short, low-trust societies have also a new degree of institutionalisation, that is political or economic organisations 'with value and stability'.[21]

In its most extreme manifestation, the cultural stream described embraces a strong notion of 'unfulfilled national aspirations'. For a fraction of the population, however marginal, there is a residual sense that the ultimate fulfilment of national aspirations should involve the relentless reaffirmation of an amorphous national ideal for the ecumenical role of Hellenism and its revival in all forms and in all places that has flourished in the past. In this context, the Hellenic heritage and the Byzantine legacy should always remind one of the 'glorious historical past', the significance and superiority of the Hellenic values which should always prevail and form the guiding principles for Greek foreign policy especially towards Turkey. Inevitably this stance precludes any compromising attitude towards Turkey, as it prevents the definitive acceptance of the fact that Hellenism is permanently confined within the borders of the present Greek state.

As a result, this marginal social group appears inclined to endorse and even amplify foreign policy objectives, however maximalist or irrational they may be. In the context of policy-making, this group assumes vastly disproportionate importance as a source of foreign policy-making. The effort on the part of foreign policy-makers to appeal electorally to this group tends to confirm upon it political significance far beyond its mere numerical size. This of course becomes possible because this group operates in a political environment within which the electorate at large is psychologically and culturally predisposed to give its support, for at least a considerable period of time before fatigue and disappointments descend, to maximalist postures in foreign policy.

## THE PERSONALITY AT THE CENTRE

The preceding analysis has sought to explain the reasons for which Greece lacks a systemic institutional framework for foreign policy-making. By systemic institutional framework, we refer to a set of interlocking policy-shaping instruments both at the political and bureaucratic levels (cabinet, committees, working groups, etc.) interacting cooperatively in such a way as to produce a policy outcome which is achievable, stable and supported by those involved in the policy-formation process, especially by the government as a collectivity.

The absence of such a framework places individual policy-makers at the centre of the process of foreign policy-making with the Minister of Foreign Affairs, or occasionally the Prime Minister, becoming the sole foreign policy-maker, operating outside the confines of any collective body of institutional structure. Our research shows that in most cases major foreign policy decisions have been taken by the Foreign Minister with the expressed or presumed consent of the Prime Minister or by the latter without prior systematic consultation, let alone extensive discussion in any collective body of policy-making, at governmental or lower political level.

A study of the various accounts of foreign policy written by former ministers of foreign affairs both under the New Democracy (ND) governments (1974–1981, 1990–1993) as well as under PASOK (1981–9, 1993–5) confirm this fundamental finding, namely that practically all important strategic foreign policy decisions were not the output of institutional procedures but the product of individual preferences. This emerges clearly from careful examination of personal accounts written, among others, by the foreign ministers or by authors whose work was openly blessed by former ministers 'as accurate'. More specifically, to support our thesis we have relied first on our personal experience and observation of how the Greek Ministry of Foreign Affairs works, and secondly, on close scrutiny of personal accounts of foreign policy-making written either by former ministers of foreign affairs, prime ministers, or in one case, by the Secretary General of the MFA. Included in these accounts are the following:

- D. Bitsios (Foreign Minister 1975–7), *Pera Apo ta Synora 1974–1977* [Beyond the Frontiers 1974–7], Athens, Estia, 1982.
- G. Rallis (Prime Minister 1980–1 and Minister of Foreign Affairs), *Ores Efthinis* [Hours of Rensponsibility], Athens, Euroekdotiki, 1983, and *Horis Kamia Prokatalipsi gia to*

*Paron kai to Mellon* [Without any Prejudice for the Present and the Future], Athens, Euroekdotiki, 1983.
- Y. P. Kapsis (Deputy Foreign Minister and Alternate Minister of Foreign Affairs), *Oi Tris Meres tou Marti* [The 3 Days of March], Athens, Nea Synora, 1990.
- M. Papakonstantinou (Minister of Foreign Affairs), *To Imerologio Enos Politikou, I Emploki ton Skopion* [The Diary of a Politician, The Skopjie Entanglement], Athens, Estia, 1994.
- G. Papoulias (Minister of Foreign Affairs and Special Negotiator for the FYROM question), 'Pos Fthasame sti Diethni Apomonosi' [How We arrived at the International Isolation], *Oikonomikos Tachydromos*, 8 November 1994.
- V. Theodoropoulos (Secretary General of the Ministry of Foreign Affairs), *Anaskopisi tis Ellinikis Eksoterikis Politikis* [Review of Greek Foreign Policy], Athens, I. Sideris/ELIAMEP, 1995.
- V. Tsouderos (Deputy Minister of Foreign Affairs), *Eksoteriki Politiki, I Megali Asthenis* [Foreign Policy, The Great Sick Man], Athens, Papazisis, n.d.

Two books which provide insight information were written by journalists, but with prologues and explicit endorsement also justify a mention:

- T. Skylakakis, *Sto Onoma tis Makedonias* [In the Name of Macedonia], Athens, Evroekdotiki, 1995 (endorsed by C. Mitsotakis).
- A.C. Tarkas, *Athina-Skopia, Piso Apo tis Klistes Portes* [Athens-Skopjie, Behind the Closed Doors], Athens, Lavyrinthos, 1995.

In the first period of the ND government (1974–81), the Prime Minister himself, Karamanlis, was the undisputed key decision-maker supported, at the technical level, by a group of close advisers, associates and friends. The Minister of Foreign Affairs, who was usually a man fully loyal and answerable to him, was amongst his advisers as were other members of the cabinet (Minister of National Defence, or Minister of National Economy). But all accounts and other research suggest that Karamanlis, a formidable charismatic yet autocratic personality, never put a question of foreign policy for in depth consideration, let alone decision, before any collective body of policy-making (cabinet, specific body). Discussion of foreign policy issues held in the cabinet served primarily to inform the members of the government and, more importantly, to give the semblance of democratic legitimacy by presenting options as collective governmental decisions. It is quite indicative in this respect that for a considerable period of time the Minister of Foreign Affairs in the Karamanlis's government was a non-political figure, an able former carrier diplomat (Bitsios) who saw his role chiefly as implementor rather than as maker of foreign policy.[22] Karamanlis's strategic foreign policy decisions concerned Greece's request to accede to the European Community, Greece's withdrawal from the military wing of NATO (August 1974) and a series of decisions concerning Greek–Turkish relations.

In the second ND government (1990–3), following a short intermission with experiments of coalition government, Samaras, a young USA-educated economist with high political ambitions and an even higher profile for a rapid political advancement but without any previous experience in foreign policy, served for two years as Minister of Foreign Affairs (April 1990–April 1992).

No doubt, the central decision associated with Samaras's tenure in the MFA concerned the Former Yugoslav Republic of Macedonia (FYROM), namely the insistence on the part of Greece that this newly-established republic, which emerged from the break-up of the federated Yugoslavia, could not and should not be recognised by the international community with the name of Macedonia or any other name derived from it.[23] The reason cited for this was that the name Macedonia along with various other elements (Constitutional provisions, state symbols, propaganda, etc.) implied expansionist designs by FYROM towards Greece. Although it is true that FYROM did pursue a policy of questioning Greece's sovereignty regarding the Greek province of Macedonia, the insistence for preventing the recognition of the Republic with the name of Macedonia landed Greece into the most adventurous course in its foreign policy pursuits in recent years leading it to almost complete isolation on this issue both in the EU and the international community at large.[24]

It is now fully documented[25] that this crucial decision was taken by Samaras himself without any previous discussion in the cabinet of the governmental council on foreign policy. In other words it was not an 'institutional decision'. It was a distinctly personal preference. Indeed this decision was taken against the advice of the diplomatic service.[26] Subsequently of course and while Greece had embarked upon the diplomatic struggle, the decision was endorsed by the so-called council of the political leaders convened by the President of Republic, Karamanlis. Nevertheless the crucial point is that in the first place the pivotal decision did not result from any institutionally defined policy-making process.

In the first period of PASOK government (1981–9) the key policy-maker in foreign policy was Andreas Papandreou, the Prime Minister. His two Foreign Ministers at that period were Haralambopoulos (1981–5) and Papoulias (1985–9), both very trusted and loyal figures to him. Papandreou, an even more charismatic personality than Karamanlis, with immense communicating skills and rhetorical capacity, was averse to submitting his authority to any institutional discipline or collective body of policy-making. As it has been observed, the cabinet, which used to meet very sporadically under Papandreou, served only to bolster Papandreou's authority by endorsing unquestionably whatever decisions he was taking.[27]

Foreign policy was indeed the policy area that Papandreou kept firmly under his personal authority, especially as he viewed himself as world statesman and was prone to speak on any major international political issue. Important foreign policy decisions falling outside the day-to-day management of policy could be made and were actually made only by Papandreou. The Minister of Foreign Affairs was usually consulted, in informal meetings that were organised with close friends or associates; but rarely, if ever, were the formal organs of foreign policy (Cabinet, Government Council on Foreign and Defence/KYSEA) convened in order to discuss major political topics of foreign policy with the view of deciding on them.

Thus the major political decisions on foreign policy: (i) the agreement concerning the USA military bases in Greece (1983), (ii) the 'Initiative of the Six' for promoting nuclear disarmament, and (iii) the decision to launch a policy rapprochement with Turkey (in 1988), were all decisions not institutionally rooted in any body. They were decisions made by Papandreou himself.

The way Papandreou manipulated the formal bodies of policy-making is eloquently illustrated by the account of Kapsis, the former Alternate Minister of Foreign Affairs (1982–9). He mentions several times that the Government Council on Foreign Policy and Defence (KYSEA) discussed extensively foreign policy issues with the aim of formulating strategy and yet the strategy actually implemented was diametrically different (or the reverse) from that 'agreed in the Council'. It is quite characteristic that Kapsis, who was in charge of Greek–Turkish relations, expresses his profound amazement 'as to why and who persuaded Papandreou to ask for negotiations with Turkey after the successful resolution of the 1987 crisis which had brought the countries to the brink of war'.[29] Speaking recently, Pangalos said that 'We have not yet overcome the bad practice of the past in making policy; only recently have we started to discover the significance of making collective decisions'.[30]

## FOREIGN POLICY IN CONFLICT

The fact that major foreign policy decisions are not 'institutional outputs' but constitute personal preferences formulated outside of any systemic institutional framework means that, by and large, the decisions are taken on the basis of the criteria reflecting the narrow options and interests of the personality involved. The paramount criterion identified in this respect is that of the maximisation of the electoral appeal and support. This by no means implies that decisions necessarily contradict the 'national interest'. On the contrary, they may appear on the surface to serve in the most pronounced fashion the national interest. This is because, by using as the guiding criterion in policy-making the broadening of the electoral appeal, the decision-maker formulates overly maximalist decisions echoing the cultural proclivities of the electorate. Indeed, one of the fundamental features associated with foreign policy decisions is that they appear maximalist and irrational in the wider context of international politics. More so, as the decisions and generally the stance decision-makers adopt are presented to the wider international community as totally not 'negotiable', that is, not subject to any compromise or trade-offs. This unavoidably ushers in very rigid postures, which usually prevents Greece from entering in time into a negotiating process capable of resolving the issues at the highest possible threshold. This results in a number of consequences:

(a) It creates a crisis atmosphere in policy-making. It is interesting to note in this context that because foreign policy takes on a crisis dimension, it tends to reinforce the non-institutional personalistic character of the foreign policy-making. In other words, under a crisis situation, the Minister of Foreign Affairs or the Prime Minister make crucial decisions with a very limited participation of other actors, normally a group of close associates. This confirms what Hill has rightly observed, namely that 'a crisis atmosphere leads to decision-making being centred on small and ad hoc groups which may reflect the sympathies of top leadership more than those of the whole governing cabinet or party.'[31]

(b) It turns the policy into a hostage to the vagaries of the electorate. This is the phenomenon known in Greece as the 'laikismos' (populism) in foreign policy, in the sense that foreign policy objectives are set primarily with the electorate in mind rather than the national interest.[32] This phenomenon accounts for the criticism frequently voiced in Greece that foreign policy is hostage to the 'political cost'. It implies that, because of the political consequences in terms of electoral support, some decisions that may have vital political choices are not taken, thereby perpetuating a stalemate in foreign policy detrimental to the national interests. The problem of 'FYROM' is frequently cited as a typical example in this respect. Although political decision-makers did realize that Greece's demand for a complete elimination of the term 'Macedonia' by the newly-established republic, was untenable yet they persisted on this course, for a considerable period of time, thereby alienating Greece from the international community.

(c) There follows instability and discontinuity in policy-making. Precisely because they are the product of personal preferences and not of institutional procedures, foreign policy choices and decisions lack stability and continuity. The change in the personalities involved in policy-making spells, more often than not, abrupt changes in the foreign policy pursuits, strategies and objectives.

Moreover, the setting of maximalistic largely irrational foreign policy objectives under the influence of electoral considerations leads Greek foreign policy into conflict with:

(i) the European Union,
(ii) the international community at large,
(iii) the advanced economic interests operating within the Greek society,
(iv) sections of the governing party and elite.

It can be easily shown that, over the past years, a host of major foreign policy pursuits have brought Greece into direct conflict with its partners in the European Union. These include: Greek policy towards the Middle East in the early 1980s, East–West relations, Greek–Turkish relations, and more specifically the decision to block the conclusions of the financial protocol between EU and Turkey, the 'FYROM question', and relations with Albania (1993–4).

The same happened with regard to international community at large.[33] A look at the comments of the international press show the extent of Greece's isolation at times of high foreign policy activity. The negative, occasionally hostile, reactions of the European Union and the international community to Greece's foreign policy objectives are usually perceived by the wider public at large as a betrayal of Greece by friends and foes alike, as a conspiracy against Greek interests as 'anti-Hellenic acts', perpetrated by 'ungrateful allies'.

Furthermore, the maximalist foreign policy choices bring Greek foreign policy into conflict with the most advanced economic interests in the country, especially with economic units which operate on rational principles aiming at maximising economic benefits. The reasons are obvious. The crisis atmosphere generated by maximalist foreign policy decisions undermines confidence, an important factor for investment. Economic activity in general and tourism in particular are adversely affected. Not suprisingly, therefore, those economic interests oppose the 'maximalist foreign policy' objectives.

Thus, the Confederation of Greek Industries (SEV) openly criticised Greece's policy towards Balkan states in the period 1991–5 and more specifically Greece's initial unyielding stance towards FYROM, Albania and Bulgaria. Indeed, it seems that Greek firms flouted the trade embargo Greece imposed on FYROM in February 1994. The overall Greek policy towards the Balkans pursued in the period 1992–4 encountered the disapproval of the business community, which saw the prospects of penetrating economically the Balkan hinterland frustrated. This was openly expressed by various organised economic interests, including the powerful Confederation of Greek Industries, the Chamber of Commerce, and the Confederation of Industries of Northern Greece.[34]

Uneasiness is also expressed, albeit in a moderate fashion, at policy measures leading to tension in Greek–Turkish relations. Heightened tension is seen as an obstacle to taking advantage of the economic opportunities offered by this country on the one hand and to increase investment activity on the other. It is quite characteristic in this connection that at the height of the tension in Greek–Turkish relations in 1996, in the wake of the Imia crisis (January 1996), and very much to the amazement of the government and the political elites in general, Greek and Turkish tourist operators signed in Athens a 'protocol of cooperation' in this vital area of economic activity providing for joint actions, joint advertisement campaigns, joint 'packages'.[35] In addition, Greek tourist operators undertook to press the Greek Ministry of Foreign Affairs to lift the visa requirements for Turkish citizens visiting Greece while both Greek and Turkish tourist agencies 'pledged to take measures with the aim of halting actions leading to sharpening tension between Greece and Turkey'. Never in the past had Greek economic interests been so openly at odds with the official policy pursued by the country over such a sensitive area (Greek–Turkish relations).

Foreign policy objectives have rarely, if ever, been endorsed by the entire government, let alone by the governing party as a whole. Foreign policy constitutes a constant source of dissent within the government irrespective of its political colour. Only under Karamanlis's government (1974–80) and only for a very limited set of issues, including the request to accede to the European Community, was there strong governmental consensus in favour of foreign policy objectives. Under PASOK's governments (1981–9) there was constant and open friction among cabinet members over foreign policy. The phenomenon took a frightening dimension during the ND government of 1990–3, when Prime Minister Mitsotakis and a number of other ministers were openly hostile to foreign policy pursued by the Foreign Minister of the same government, Samaras, particularly in relation to the Balkans. This led eventually to the dismissal of Samaras as Foreign Minister (April 1992) and the assumption of the post by Prime Minister Mitsotakis himself and subsequently by Papakonstantinou. When this team inaugurated a conciliatory policy towards FYROM, Samaras and his followers in parliament threatened to withdraw their support from the government. The collapse of the Mitsotakis government occurred in September 1993 because of Samaras's defection.

The phenomenon of government's disarray over foreign policy repeated itself under the PASOK government which took office in October 1993. There was even dissension within the Ministry of Foreign Affairs itself: Pangalos, the Minister for European Affairs, was openly opposed to the policy pursued by the Foreign Minister, Papoulias, towards the Balkans (FYROM) and other areas. Pangalos, for instance, expressed himself against the imposition of the trade embargo on FYROM in February 1994, in the middle of the Greek Presidency. Predictably, the decision for the embargo had not been taken by a collective institutional body of policy-making.[36]

The ensuing isolation of Greece in the European Union and the international community at large, usually leads, after a period of time, to the process of searching for a compromise for settling the issue concerned. The important point is however that this quest for compromise starts from a point of greater weakness for Greece owing first of all to its isolation within the international enviroment and the loss of support even from its EU partners.

Two important cases illustrate this point. First, as was noted above, the FYROM question. In April 1992, Greece at the height of its policy intransigence turned down a package of measures assembled by the European Commission (Pinheiro package) for resolving the issue. This package was thought to be a reasonable compromise favourable to the Greek views but not a wholesale vindication of the Greek view concerning the name of the new republic. For this reason, Greece rejected the proposal, leading the EC to abandon any attempt on its part at brokering a compromise. Almost four years later, in October 1995, Greece accepted an 'interim agreement' with FYROM worked out under the auspicies of the UN special mediator, Vance. Though highly appreciated, this agreement was generally considered as more remote to the Greek pursuits compared to that offered by Pinheiro. Moreover, it did not tackle the question of the name, which by that time had been widely accepted as being 'Macedonia'. In other words, Greece was forced to compromise at a much lower threshold than it could have been the case had it accepted the Pinheiro proposals.[37]

The second case concerns Greek–Turkish relations. In 1986 Greece put a veto on the conclusion of the fourth financial protocol between the EC and Turkey. Greece set as a condition for lifting its veto the withdrawal of the Turkish military forces from the occupied part of Cyprus. Although isolated on this issue within the EC, Greece insisted on this posture until Autumn 1994 when it was compelled to abandon this condition. According to the new situation, the EU was asked to open negotiations for the accession of Cyprus to the EU six months after the end of the Intergovernmental Conference of 1996 for the revision of the Maastricht Treaty. A statement to this effect was offered by the EU Council, and accordingly Greece lifted its nine-year veto in March 1995.

A good number of former diplomats and ministers of foreign affairs have described Greek foreign policy concerning vital issues including the 'Cyprus problem', Greek–Turkish relations, and relations with the Balkans 'as a sequence of successive mistakes and lost opportunities'. Ambassador Theodoropoulos, former Secretary-General of the Ministry of Foreign Affairs, has pointed out that: 'the maximalistic, largely irrational objectives and the uncompromising, unyielding stance in the foreign policy domain have resulted either in national catastrophies and defeats or, at best, in the acceptance of worse solutions than we could have achieved through a more flexible strategy'.[38]

## CHALLENGES TO GREEK FOREIGN POLICY

The preceding analysis, which has sought to identify the model of foreign policy-making in Greece and the fundamental weaknesses associated with this model, allows us to seek out the basic challenges which foreign policy formation needs to address in order to forge a systemic institutional framework for elaborating foreign policy and making decisions of an institutional nature. In our view, these challenges are:

- (i) Institutionalisation
- (ii) Europeanisation
- (iii) Economisation
- (iv) De-externalisation

### (i) Institutionalisation

As our analysis has shown, the establishment of a systemic institutional framework for foreign policy-making constitutes the indispensable condition for reshaping Greece's model of foreign policy-making. True, the building of stable, effective, legitimate and workable institutions and procedures presupposes far-reaching changes in the cultural environment that prevails in Greece. In so far as these changes are taking place, and some changes are indeed taking place, in the form of the emergence of modernising culture based on rational features, a systemic institutional framework can be built. This after all is now presented as a main prerequisite by all political parties for the improvement of the effectiveness of Greece's foreign policy.[39]

Thus New Democracy proposes the creation of a permanent post of Deputy Minister of Foreign Affairs as a means of ensuring stability and continuity in policy-making.[40] The Alliance of the Left (SYN) has long argued for the setting up of a National Council of Foreign Policy composed of representatives of all political parties 'as an instrument of ensuring rationality, effectiveness, continuity and stability in foreign policy'.[41] A similar proposal has also been put forward by ND. Generally, all political parties from right to left insist that the building of more stable institutional structures holds the key to improving the foreign policy outputs and coherence and effectiveness of foreign policy.

The new PASOK government under Simitis, which took office in January 1996 following the resignation of Andreas Papandreou, marked an important stage in the struggle for the modernisation of Greece's economic and political system in line with the requirements stemming from the EU membership. A fundamental element of this process is the strengthening of the collective character of the policy-making process. In this context, an attempt is being made to turn the so-called Kyvernitikon Symvoulion (Government Council-Cabinet) into the central policy-formulating and decision-making body.[42] To that effect, a new secretariat and other supporting units modelled on the organisation of the European Commission have been set up, headed by technocratically educated personnel.[43]

As part of this overall design, upon assuming office, the new Minister of Foreign Affairs, Pangalos, with his extensive knowledge of European affairs, due to his term as Alternate Minister of Foreign Affairs, has inaugurated a systematic effort at reorganising the Ministry with the stated aim 'of transforming it into a modern instrument of foreign policy-making'. In this connection, a committee of experts was set up to revise the internal statute for the organisation of the Ministry, a committee of eminent experts in international law was created to assist in foreign policy-making and the establishment of a new policy-planning department is envisaged. In general, Pangalos seems to attach overriding importance to modernising the structure of the Ministry including the diplomatic service. In addition to all other measures planned, a system of constant training and education of diplomats in modern subjects of diplomacy is to be introduced.[44]

Given Pangalos' determination, it is widely thought that he is uniquely qualified to introduce the reforms long overdue in the Greek MFA. Yet, what might defeat this effort is the impulsive personality of Pangalos himself, which in a way resists the limitations associated with a systemic institutional framework of foreign policy-making.

From a broader perspective, however, the progress in the process for building an institutional system of policy-making in Greece hinges on the ability of the political elites and the country as a whole to remould deeply-entrenched cultural and societal traits, something which is bound to be a long-term process.[45]

## (ii) Europeanisation

Membership of the European Union is seen as a very powerful parameter in shaping a country's economic, social and political dynamics including foreign policy objectives and management. Especially after the establishment of the European Political Cooperation (EPC) and more significantly after the putting into effect of the Treaty on European Union (TEU), with its provisions for a Common Foreign and Security policy (CFSP), the foreign policy objectives of member states cannot but be placed within the wider context of the EU even if these objectives remain essentially 'national'. Moreover, the scope of foreign policy as well as its conduct change in a profound way as a result of EU membership.[46]

The scope of foreign policy is vastly enlarged because a small country comes, as a member of the EU, to be confronted with a wide range of issues going well beyond the narrow agenda of the topics a small country usually confronts outside the EU. By virtue of its membership of the EU, a small country has the opportunity to participate in a large number of groups in the EU and in international bodies ranging from GATT (WTO) to the UN and present either EU views or national views consistent with EU policy. This role is immensely reinforced whenever a country holds the Presidency of the Council of the EU.

In a similar way the conduct of a country's foreign policy is thoroughly transformed as a result of the country's membership of the EU. The communication system changes, the representation to the outside world also changes considerably, the model of external contacts changes too, due to almost daily meetings either in the context of the Political Committee or in other groups in Brussels.[47]

In short, membership of the EU dictates that foreign policy should shed its parochial narrow national character and acquire a more global dimension at least as regards its scope and agenda. Moreover, it demands a regularisation in the conduct of policy, entailing stability, continuity, periodicality in meetings. In a nutshell, it involves greater institutionalisation in foreign policy-making and management.

Although Greece has recently made some progress in adjusting its foreign policy to the EU requirements, its foreign policy is still disproportionally overburdened by the so-called 'national issues', notably relations with Turkey, Cyprus and, most recently, 'the FYROM question'.[48] Most worryingly, it tends more often than not to filter all other European or international issues through the Greek–Turkish prism. This of course renders Greek behaviour and performance in the EU ineffective as it appears to be constantly preoccupied with a limited and quite predictable range of issues. Coupled with the institutional malfunctionalities, this narrowly oriented approach in defining the country's foreign interests and in conditioning negotiating behaviour in the context of CFSP generates the impression that Greek foreign policy cannot adjust to the EU requirements. No matter how exaggerated or misplaced this perception may be, at least in certain cases, it certainly weakens Greece's position in the EU. To overcome this, Greek foreign policy should be adjusted to requirements of EU membership; it should be Europeanised in the manner we described above.

### (iii) Economisation

It has been widely recognised that economic issues (economic diplomacy) have increasingly come to occupy a very important place in the foreign policy of all states, large or small. This is due to a host of factors including the deepening of economic interdependence, the globalisation of the economy and the communication system, technological developments, etc.[49] In the context of European integration, it has long been observed that 'the sharing of functions between national and Community institutions blurs the distinctions traditionally made between domestic and foreign policy'.[50]

Indeed, economic issues ranging from agriculture to macroeconomic and monetary policy, environment, trade, research and development, structural policy or what has been termed 'low politics' constitute the hard core of the negotiating agenda that the diplomatic service confronts within the context of the EU. Furthermore, in their relations with the outside world, in the wider international system, the diplomatic services of member states of the EU are forced to deal with complex economic issues of technical nature. All these developments make the 20th century diplomacy 'total in its objectives and subject matter'.[51] For a small country, especially a European country (such as Belgium, the Netherlands, etc.) economic issues in the broader sense form the basic stuff of their foreign policy.

In Greece, the supremacy of high politics in foreign policy embracing issues like relations with Turkey, the Cyprus question, relations with the superpowers, particularly the USA, has deprived the Ministry of Foreign Affairs of the opportunity and will to develop skills and mentality for dealing with economic issues. Indeed until recently, the Greek diplomatic service manifested a clear distaste of handling economic and technical issues. This was the reason for which a new service composed of experts was set up in the Ministry in 1976 to deal with specialised economic and technical issues.[52] Membership of the European Union and the ensuing need to shape and coordinate policy on a wide range of economic issues put severe strains on the Ministry.

The need therefore to strengthen the capacity of the Ministry to handle specialised economic issues has become obvious. Not only membership of the EU but the overall developments in the regional and international system dictate that. Relations with Greece's Balkan neighbours have, for instance, assumed recently a strong economic dimension. Pangalos, the current Minister of Foreign Affairs, places high emphasis on the economic aspects of foreign policy. He announced recently that the percentage by which a diplomat increases Greece's exports would be a criterion for promotion.[53]

### (iv) De-externalisation

The blurring of functions between foreign policy and domestic policy resulting from the EU membership implies that other ministries than the Ministry of Foreign Affairs come to play an important role in the foreign policy domain. Likewise, other bodies and institutions of the society including academic and research centres, local and regional administration have, as a result of EU membership and other more general factors (interdependence, etc), assumed significant capabilities to influence the foreign policy and external relations.

More importantly, perhaps, the economic agents (advanced and export-oriented firms and enterprises) have emerged as a key factor in the foreign policy field by developing economic links with the neighbouring Balkan countries. As noted, in spite of the tension that was associated with Greece's official policy towards the Balkans in the period 1991–5, Greek firms made spectacular inroads into the Balkan countries by penetrating their markets in terms of exports and investment.

The cumulative effect of these developments signify, in broad terms, a process which can be described as the 'de-externalisation' of foreign policy. In other words, the Ministry of Foreign Affairs has become just one, albeit the supreme one, body of making and managing policy. However its 'monopoly' in foreign policy has been circumscribed by the growth of alternative bodies, be economic, social, scientific or academic, sharing in foreign policy functions.

This phenomenon demands further adjustments in the system of foreign policy in order to ensure a degree of consistency and coordination in the foreign policy outputs and activities at least at the state level. One of the persistently negative features of Greece's European policy has been the lack of coordination of policy presented to the European Union or other international fora.[54]

## CONCLUDING REMARKS

The study of the process of foreign policy-making in Greece reveals the underlying significance of personalities and society as 'sources' of foreign policy in the absence of systemic institutional framework of foreign-policy-making. It also reveals how these 'sources' determine the content, objectives, features and 'quality' of foreign policy, if the latter is made in a domestic institutional vacuum. Moreover it shows how the external environment (European Union, regional and international system) can generate strong pressures for the adjustment and reorientation of a country's foreign policy. And yet in the absence of an institutional framework to absorb and 'process' these pressures, the system fails to adjust quickly and smoothly to new environmental conditions and requirements. From a broader theoretical perspective, the analysis of Greek foreign policy, while underscoring the significance of personalities, underlies the pivotal importance that institutions have in producing rational foreign policy choices as well as ensuring effectiveness in foreign policy.

## NOTES AND REFERENCES

1. For comments on Greece's foreign policy by non-Greek commentators see, among others: S. Talbot 'Greece's Defence Seems Just Silly', *Time* 12 October 1992 'Stupid or Evil?', *The Spectator*, 9 April 1994; M. Simons 'Club Europe's Private Doubts: Greece Won't Play by the Rules', *International Herald Tribune*, 9 April 1991; 'The Sick Man of Europe', *The Economist*, 9 May 1992; M. Mazower 'Classic Errors in the Balkans', *Guardian*, 12 April 1994; 'L'Europe Sans le Sud', *Le Monde*, 7 May 1994; 'Athens on Trial', *The Times*, 8 April 1994.
2. C. Mitsotakis, 'Prologos' in T. Skylakakis, *Sto Onoma tis Makedonias* [In the Name of Macedonia], Evroekdotiki, Athens, 1995, p. 2.
3. C. Simitis, *Yia Mia Koinonia Ischiri, Yia Mia Ichiri Ellada* [For a Strong Society and a Strong Greece], Plethron, Athens, 1995, p.157.
4. For an academic/professional analysis of Greece's foreign policy among others see V. Theodoropoulos, E. Lagakos, C. Papoulias, J. Tzounis, *Skepseis ke Provlimatismoi yia tin Exoteriki Mas Politiki* [Thoughts and Reflections on Our Foreign Policy], I. Sideris/ELIAMEP, Athens 1994; D. Konstas, P. I. Tsakonas, Elliniki Exoteriki Politiki [Greek Foreign Policy], Odysseas, Athens, 1994; V. Theodoropoulos, *Anaskopisi, I Exoteriki Politiki tis Neoteris Elladas* [Review, the Foreign Policy of Modern Greece], I. Sideris/ELIAMEP, Athens 1995; C. L. Rozakis, *I Exoteriki Politiki ke oi Evropaikes Koinotites* [Greek Foreign Policy and the European Communities], Foundation of Mediterranean Studies, Athens 1988; T. Veremis, T. A. Couloumbis, *Elliniki Exoteriki Politiki Prooptikes ke Provlimatismoi* [Greek Foreign Policy: Prospects and

Reflections], I. Sideris/ELIAMEP, Athens 1994; T.A. Couloumbis, 'Introduction: The Impact of EC(EU) Membership on Greece's Foreign Policy Profile' in P. Kazakos and P. C. Ioakimidis (eds), *Greece and EC Membership Evaluated*, Pinter, London, 1994; T. A. Couloumbis, P. Yannas, 'Greek Foreign Policy Priorities for the 1990s', in K. Featherstone and K. Ifantis (eds), *Greece in Changing Europe*, Manchester, University Press, Manchester 1996; P. C. Ioakimidis, 'Contradiction in the Europeanisation Process', in Featherstone, Ifantis, op. cit.; Y. Valinakis, *Greece's Security in the Post- Cold War Era*, Stiftung Wissenschaft Politici Ebenhausen, 1994.

5. C. Hill, 'A Theoretical Introduction', in W. Wallace and W. E. Peterson (eds), *Foreign Policy-Making in Western Europe, A Comparative Approach*, Saxon House, London, 1978.
6. M. Clarke, 'Foreign Policy Analysis: A Theoretical Guide', in S. Stavridis, C. Hill (eds) *Domestic Sources of Foreign Policy, West European Reactions to the Falklands Conflict*, Berg, Oxford, 1996.
7. C. Hill, 'A Theoretical…', op. cit., p. 21.
8. M. Clarke, 'Foreign Policy…', op. cit., p. 22.
9. On this point see, A. Makrydimitris, *H Dioikisi se Krisi, Keimena yia ti Dioikisi ke tin Koinonia* [The Administration in Crisis, Texts for the Administration and Society], Nea Synora/Ant. Livanis, Athens 1996; D. A. Sotiropoulos, *Grafeiokratia ke Politiki Exousia* [Bureaucracy and Political Power], Ant. Sakkoulas, Athens 1996; R. Clogg (ed.), *Greece 1981–1990, the Populist Decade*, Macmillan, London 1993; H. J. Psomiades and S. B. Thomadakis (eds), *Greece, the New Europe and the Changing International Order*, Pella, New York, 1993.
10. Clogg, '*Greece*…', op. cit.; T.C. Kariotis (ed.), The Greek Socialist Experiment, Papandreou's Case *1987-1989*, Pella, New York, 1992.
11. On the organisation of the Ministry of Foreign Affairs see T. Stoforopoulos, A. Makridimitris, *To Systima tis Ellinikis Exoterikis Politikis, I Thesmiki Diastasi* [The Greek System of Foreign Policy, the Institutional Dimension], EKEM Research Paper no. 38, Athens, 1996.
12. On the role of diplomacy see K. Hamilton and R. Langhorne, *The Practice of Diplomacy: Its Evolution, Theory and Administration, Routledge, London, 1995.*
13. Personal interviews and accounts.
14. Among those who resigned are ambassadors Doundas and Stoforopoulos over Greece's decision in Davos in 1987 to 'drop' Cyprus as a prerequisite for improving Greek–Turkish relations.

15. Personal interview.
16. On this see T. Veremis, 'From the Nation State to the Stateless Nation 1821–1910', in M. Blinkhom, T. Veremis, *Modern Greece: Nationalism amd Nationality*, Sage/ELIAMEP, Athens, 1990.
17. P.N. Diamantouros, 'Politics and Culture in Greece; 1974-1991: An Introduction', in Clogg, '*Greece* . . . ', op. cit., pp. 1–25.
18. N.P. Mouzelis, 'Greece in the Twenty-first Century: Institutions and Political Culture', in D. Constas and T. C. Stavrou (eds), *Greece Prepares for the Twenty-first Century*, Woodrow Wilson Centre Press/John Hopkins University Press, Washington DC, 1995.
19. On this phenomenon see also K.R. Legg, *Politics in Modern Greece*, Stanford University Press, Stanford, Cal., 1969.
20. F. Fukuyama, *Trust, the Social Virtues and the Creation of Prosperity*, Free Press, London, 1995, pp.5–21.
21. Ibid.
22. See D. Bitsios, *Pera apo ta Synora 1974–1977 [Beyond the Frontiers 1974–1977], Estia, Athens, 1982.*
23. In addition to references already mentioned, see also Y. Valinakis, S. Dalis (eds), *To Zitima ton Skopion* [The Skopje Question], I. Sideris/ELIAMEP, Athens 1994.
24. See M. Papakonstantinou, 'Na Termatisoume tin Lampri Apomonosi' ['Let's End our Glorious Isolation'], *Anti*, no. 601, 16 March 1996; C. Papoulias, 'Pos Fthasame stin Diethni Apomonosi' ['How We Arrived at the International Isolation'], *Oikonomikos Tachydromos*, 3 November 1994.
25. See M. Papakonstantinou, *To Imerologio Enos Politikou* [The Diary of a Politician], Estia, Athens 1994; T. Skylakakis, *Sto Onoma tis Makedonias* [In the Name of Macedonia], Evroekdotiki, Athens, 1995; A. C. Tarkas, *Athena–Skopia, Piso Apo tis Kleistes Portes* [Athens–Skopje, Behind the Closed Doors], Lavyrinthos, Athens, 1995.
26. Personal interview with the Political Director in the MFA, C. Lymberopoulos.
27. Pangalos, Minister of Foreign Affairs, Interview, *Eleftherotypia*, 19 May 1996.
28. Y. Kapsis, *Oi 3 Meres tou Marti* [The Three Days of March], Nea Synora/Ant. Livanis, Athens, 1990.
29 Ibid., p. 120.
30 T. Pangalos, Interview, op.cit.
31 C. Hill, 'A Theoretical . . . ', op.cit., p. 12.
32 P. Kazakos, 'Paradosi ke Laikismos stin exoteriki Politiki' [Tradition and Populism in Foreign Policy], *Eleftherotypia*, 9 May 1994.

33. On this see Y. Valinakis, 'I Ellada stin Evropaiki Politiki Synergasia, ta Deka Prota Chronia' [Greece in the European Political Cooperation: the First Ten Years], in L. Tsoukalis (ed.), *Ellada stin Evropaiki Kinotita, I Proklisi tis Prosarmogis* [Greece in the European Community, the Challenge of Adjustment], EKEM/Papazisis, Athens, 1993.
34. See S. Papaspiliopoulos (ed.), *Ta Balkania stin Enomeni Evropi, I Symboli tis Elladas* [The Balkans in the United Europe: the Contribution of Greece], I. Sideris, Athens, 1995.
35. *Kathimerini*, 9 May 1996.
36. P.C. Ioakimidis, 'The Role of Greece in the Development of EC Mediterranean Policy', in R. Gillespie (ed.), *Mediterranean Politics*, vol. 2, Pinter, London, 1996.
37. C. Rozakis, *Politikes ke Nomikes Diastaseis tis Metavatikis Symfonias tis Neas Yorkis Metaxi Elladas ke FYROM* [Political and Legal Aspects of Interim Agreement of New York Between Greece and FYROM], I. Sideris/ELIAMEP, Athens, 1996.
38. V. Theodoropoulos, *Rnaiskopisi*..., op. cit. in text above...p.135.
39. See Stoforopoulos, Makrydimitris, *To Systima*..., op. cit.
40. Ibid, pp. 54–55.
41. Ibid, p. 56, Also article by Synaspismos, 'Exoteriki Politiki' [Foreign Policy], *Avgi*, 19 May 1996.
42. In May 1996 Simitis pledged that foreign policy would always be shaped in the future within the framework of KYSEA (Kyvernitikon Symvoulion Exoterikis Politikis Amynas – Government Council of Foreign Defence Policy), *Kathimerini*, 25 May 1996.
43. On the detail of this see *Eleftherotypia*, 7 May 1996.
44. Y. Kartalis, 'I Nea Poreia sta Exoterika' [A New Course in Foreign Policy], *To Vima*, 17 March 1996.
45. J. Peffifer, 'Greek Political Culture and Foreign Policy', in Featherstone and Ifantis, *Greece in changing Europe*, op. cit., pp. 17–23.
46. On the CSFP see S. Stavridis, *Looking Back to See Forward: Assessing the CFSP in the light of EPC*, European Institute, LSE, 1994.
47. P.C. Ioakimidis, 'Greece in the EC: Policies, Experiences and Prospects', in H. J. Psomiades, S. B. Thomadakis, Greece, the New Europe. Also, P.C. Ioakimidis, 'I Elliniki Dioikisi ke I Diamorfosi tis Evropaikis Politikis' [The Greek Administration and the Formulation of European Policy], in Tsoukalis, *Ellada*..., op. cit., pp.209–30; A. Makridimitris and A. Passas, *I Elliniki Dioikisi ke o Syntonismos tis Evropaikis Politikis* [The

Greek Administration and the Formulation of European Policy], Ant. Sakkoulas, Athens, 1994.

48. P. C. Ioakimidis, 'The Role of Greece in the Development of the EC Mediterranean Policy', in R. Gillespie (ed.), *Mediterranean Politics*, vol. 2, Pinter, London, 1996, p. 67–81.
49. Hamilton and Langhorne, *The Practice* . . . , op. cit., pp. 183–227.
50. H. Wallace, *National Governments and the European Communities*, Chatham House: PEP, London 1973.
51. Hamilton and Langhorne, *The Practice* . . . , op. cit., p. 183.
52. E. Kakoulidou, 'Adranes Yliko I Elliniki Exoteriki Oikonomiki Politiki' [The Non-existent Greek Foreign Economic Policy], *Kathimerini*, 5 May 1996.
53. T. Pangalos, speech, Athens 17 March 1996.
54. Ioakimidis, 'I Elliniki Dioikisi...', op. cit.

# 7 Cyprus Foreign and Security Policy: Options and Challenges

Charalambos Tsardanidis
and Yannis Nicolau

## INTRODUCTION

Cyprus' independence was realised through the so-called Zurich Agreement, which was reached on a draft plan for a settlement on 11 February 1959 between Greece and Turkey. Subsequently, the agreement which was signed in London between Greece, Turkey, Britain and the representatives of the two communities was supposed to provide a final settlement of the issue.

However, Cyprus was invaded by Turkish forces on 20 July 1974. Ankara used as a pretext the coup d' état engineered by the Athens military regime of 15 July 1974 against Archbishop Makarios and based its arguments on the Treaty of Guarantee which was signed on 19 February 1959.[1] But Turkey, instead of securing the island's independence and unity, managed to occupy nearly 40 per cent of the Cyprus Republic' s territory, thus achieving its long established goal: the partition of Cyprus.

Since 1974 an essential factor for the security of the Republic of Cyprus, on an international level, lies in the close relationship between the geopolitical situation of the island and the strategic framework of the Eastern Mediterranean. Cyprus is situated in a region exposed to conflicts spreading from the Middle East to the Persian Gulf. When viewing the problem of Cyprus in general, and particularly in the present circumstances, one must not forget the vital geopolitical importance of the island in the Eastern Mediterranean.

Cyprus although small in size, has played an important role in regional politics ever since its independence in 1960. Its strategic position, not only in terms of geography but also vis-à-vis relations with states having different worldviews, has given Cyprus the opportunity to execute functions far beyond its actual dimensions.

This chapter is divided into two parts: the first analyses critically the options of the island's foreign and security policies and the second treats separately the main challenge to Cyprus' foreign policy in the 1990s: its admission to the European Union.

## CYPRUS FOREIGN AND SECURITY POLICY OPTIONS SINCE 1974

Since 1974 Cyprus has focussed most of its foreign policy efforts on winning broader international support for a settlement providing for a withdrawal of Turkish occupation troops. Its objective has been to pursue a most favourable constitutional and territorial settlement and, to the extent possible, a restoration of the *status quo ante*.

In order to pursue these general, although essential, foreign policy objectives Cyprus has pursued since 1974 a multifaceted approach which contains six parallel strategies: intercommunal talks, internationalisation, building a special relationship with Greece, seeking American mediation, the development of a credible deterrence strategy and admission to the EU.

Each one of these strategies is fundamentally complementary to its fellow strategies although emphasis on each of them has shifted according to the prevailing international and domestic circumstances. Thus, during the periods 1975–83 and 1988–93 the emphasis has been on the intercommunal talks, while during the period 1983–8 the internationalisation strategy had obtained a priority. Since 1993 the strategy of the admission to the EU seems to have attracted the attention of the Cypriot foreign policy establishment, in conjuction with the strategy of building a credible deterrence in cooperation with Greece. On 2 March 1993, the foreign minister of the Cypriot government, Michailides, stated 'The main axis of our foreign policy is what we call our European orientation. By this, we do not only mean the promotion of our application for accession to the European Union and the harmonisation of the structures existing in Cyprus with those of Europe, in order to become a full member of the European Union, but also the activation of the European factor in the efforts to find a solution to the Cyprus problem.'

**The Intercommunal talks strategy**

The main objective of this strategy was for the Greek Cypriots to try to reach a political accommodation with the Turkish Cypriot community. The initial policy position regarding the intercommunal talks was to return to the pre-1974 situation, namely to continue the dialogue between Clerides and Denktash by granting greater local autonomy to the Turkish Cypriots. Neverthless, the December 1974 Athens meeting between Cypriot President Makarios and Greek Prime Minister Karamanlis laid down a more pragmatic strategy based on the principle of a bi-zonal and bi-communal federation.[2] These major concessions by the Greek side were included in a communique issued on 12 February 1977, at the end of a UN-sponsored meeting between President Makarios and Turkish Cypriot leader Rauf Denktash.

Since 1975, there have been several sets of negotiations and related initiatives. Despite setbacks, discussions continue to focus on ways to establish a new constitutional arrangement for the state of Cyprus that will regulate the relations of the two communities on a federal, bi-communal, and bi-zonal basis. The fundamental question that has dominated all intercommunal talks is how to restructure Cyprus constitutionally into a viable and functional federal, bi- communal system and territorially into a bi-zonal system, while maintaining the unity, sovereignty, and territorial integrity of the Republic and protecting the rights of all Cypriots.[3]

Turkish Cypriots systematically place emphasis on bi-zonality, security guarantees, and political equality of the two communities. Greek Cypriots emphasise the freedom of movement, of property, of settlement, and the return of occupied territory. Turkish Cypriots favour a confederation of two nearly independent states living side by side with limited contact, while the Greek Cypriots envision a more interdependent and integrated structure.

In 1992, a new impetus was given to UN initiatives seeking to reconcile the positions of the Greek and Turkish communities. The Secretary-General of the UN proposed a number of confidence-building measures which, had they been accepted by both parties, could have helped to achieve an overall agreement based on a 'set of ideas' approved by the UN Security Council, which envisaged the establishment of a bi-communal and bi-zonal federation, comprising one territory composed of two politically equal federated states. This federal republic, however, would have retained one indivisible sovereignty.[4]

Although some progress has been made in protracted negotiations between the parties, many questions remain unresolved, as Turkey's alleged support for a bi-communal, bi-zonal federation has proved to be at best lukewarm. The statements made at intervals by the Turkish and the Turkish Cypriot side, as well as the official position adopted by the latter during the intercommunal dialogue, suggest that the Turkish side aims at a confederation of two separate and independent states, and, in this way, at the legislation of the status quo. Moreover, whenever the Secretary General has submitted over the years concrete proposals leading in fact to a bi-communal, bi-zonal federation, the Turkish side has rejected them or sidestepped any progress e.g. the cases of working points (1984), integrated documents (1985), draft framework agreement (1986), procedural formula (1987), outline for an overall agreement (1988), Set of Ideas (1922), and Confidence Building Measures (1993). Another reason why the negotiations have failed is that the role of the UN Secretary-General has changed from being a provider of good offices to that of a mediator. This, in turn, has raised fears in both camps (Greek Cypriots and Turkish Cypriots) who tend to fear an externally imposed solution. Under such conditions the intercommunal talks have remained deadlocked since May 1994.

In sum, the Greek Cypriots believe that Turkey's consistent objective is to partition the island and, later, to establish a Turkish protectorate over all of Cyprus. Therefore, they argue there is no sense in the Cyprus government resuming the intercommunal dialogue. Their priorities should be directed towards building up a closer security relationship with Greece and seeking full membership in the European Union.

## The Internationalisation strategy

This strategy seeks to have Cyprus utilising its international posture to secure sanctions and other forms of pressure against Turkey in the international arena, and especially at the UN. The aim has been to amass enough diplomatic pressure on Turkey that will convince it to make concessions to Cyprus, especially by way of withdrawing its troops of occupation.[5]

The republic has continued to enjoy exclusive international recognition as the legal government of Cyprus. Cyprus' membership of the Non-Aligned Movement (NAM), the Organisation on Security and Cooperation in Europe (OSCE), the Council of Europe, the United Nations, and the Commonwealth provide opportunities to promote these aims. Resolutions passed by these organisations have called for the withdrawal of foreign troops, condemned Turkey's settler policy, urged the immediate implementation of UN resolutions, and called for sanctions against Turkey.

Cyprus has placed considerable importance on its membership of the NAM. It hosted a number of NAM meetings and headed an effort in 1989 and 1990 to redefine the NAM's objectives in the light of the dramatic changes in East–West relations and the virtual end of superpower rivalry and competition. Support from the non- aligned states was particularly important in the UN setting. The Greek Cypriot leadership was aware that UN resolutions could not have a decisive effect on Turkey unless they were accompanied by substantive sanctions, but it hoped that collective international pressure might yield some results. Especially, the Cypriot government sought to enhance the cumulative impact of UN General Assembly resolutions through stronger wording and tone.

The stategy of internationalisation became more promising in the early 1990s due to two new developments:

The first was the collapse of East Germany's communist regime. After the fall of the Berlin Wall and the commitment to unification of the two Germanies, the Cypriot government sought to remind the world that Cyprus remained divided through the force of arms. It argued that, after Soviet troops completed withdrawing from Eastern Europe, Cyprus would be the only country in Europe with foreign occupation troops. The decline, at least initially, of the relative importance of NATO among European institutions had both advantages and disadvantages for Cypriot foreign policy. On the one hand, it appeared to reduce Turkey's leverage over its Western allies and opened the way for broader pressures on Ankara. On the other hand, the potential loosening of Turkey's ties with Western partners could also weaken those countries' tolerance vis- à-vis Turkey's revisionist foreign policies. But the preoccupation with Germany and the emergence of violent conflicts in the Balkans made it harder to keep the attention of European powers focused on Cyprus.

The second development involved the Iraqi invasion of Kuwait on 2 August 1990, leading to the Gulf War, and the implementation of the United Nations resolutions on Kuwait by the Security Council. 'Unanimous decisions by the Security Council concerning the Persian Gulf were made possible by detente between the superpowers, and could also have served to set a precedent for the implementation of other Security Council resolutions on problems such as the Cyprus issue.'[6]

The Cypriot foreign policy planners sought to take advantage of the emerging 'new world order' found in the statements by President Bush and asserted that the status quo in Cyprus was also unacceptable.[7] Unfortunately, the American Administration did not find any similarities between aggression in Kuwait and Cyprus.[8]

In fact, experts representing the US Administration stated on 17 April 1991, before the Subcommittee on European Affairs of the Senate Committee on Foreign Relations that the UN resolutions on Cyprus had been adopted under Chapter VI (on peaceful settlement of disputes) of the UN Charter, whereas the Kuwait resolutions had been based on Chapter VII (on coercive action). Furthermore, American officials have argued that the Gulf war has changed Turkey's role in an enduring way. As Van Coufoudakis has argued, '[t]hey eagerly conclude that the Gulf war confirmed Turkey's role as a regional power with regional interests, a country that cannot be bogged down with minor bilateral problems as in the past'.[9]

The argument that the case of Cyprus is not comparable to that of Kuwait is questionable for two reasons: first, because

Cyprus' occupation is a threat to peace given the failure of the occupying power even to recognise the government of the republic; second, because a significant number of UN resolutions contain provisions calling for the implementation of the UN decisions on Cyprus[10] and urge all member states to comply with them.

There is no doubt, however, that the policy of internationalisation has not directly contributed to the resolution of the Cyprus problem. It was at best partially successful; securing statements of high principle but failing to produce concrete sanctions against Turkey. However, 'although UN's General Assembly resolutions have not legally binding force on states, and Turkey has not complied with those on Cyprus, the Greek Cypriots have derived great political benefits from them'.[11] The modest success of the policy of internationalisation explains why the United States and Turkey specifically sought the de-internationalisation of the Cyprus problem. The United States and Turkey preferred a solution of the Cyprus problem through intercommunal talks under United Nations auspices, on the basis of political equality, away from international fora. 'Cyprus may have had the ability to build a wide ranging international coalition, but neither Cyprus nor its allies had the political influence to rearrange the policy priorities of the influential powers of the international community'.[12]

## The strategy of building up a special relationship with Greece

After the troubles of 1963–4 and the virtual separation of the two communities, the Cypriot government did not, ironically, orient its foreign policy toward Greece. Instead, the growing authority and confidence of President Makarios, together with divergent trends in Greek and Greek-Cypriot politics, led to Cyprus' foreign policy becoming more independent. The Greek- Cypriot leadership was disappointed that Greece had placed the interests of the Western alliance above those of the island in the pre-independence London and Zurich talks. Greek Cypriots also viewed as inadequate the Greek response to the 1963–4 troubles, with Greece again deferring to NATO interests. Relations deteriorated further when the military seized power in Athens in 1967. By the early 1970s the rift between the Athens junta and the Makarios government had become open. Athens financed operations of anti-Makarios organisations and newspapers and was widely thought responsible for attempts on Makarios's life. Pressures mounted, and in July 1974, after Makarios openly challenged the junta's interference, the Cypriot National Guard, led by Greek junta officials, staged a coup that ultimately resulted in Turkish intervention and the junta's demise.

Following the 1974 restoration of democracy in Athens and the atmosphere of crisis after the Turkish invasion, relations between Nicosia and Athens were restored to a friendly, bilateral relationship and closer coordination of foreign policy began, particularly focused on winning support for resolutions in international organisations and securing assistance from Greeks abroad. Greece gave full support to policies adopted by the Cypriot government and pledged not to interfere in domestic Cypriot politics. The proclaimed policy of the government of New Democracy in the 1970s was 'Cyprus decides and Greece supports'. The two governments agreed that Greek Cypriot participation in settlement efforts was essential and tried to uncouple the Cypriot issue from other Greek–Turkish disputes. However differences remained over the two governments' priorities. Greek Prime Minister Karamanlis was said to favour a moderate and conciliatory stand on Cyprus compared to either Makarios or Kyprianou, both of whom advocated a ' long struggle'. The Greek government was also eager to return to NATO, (it had withdrawn from its military structure following the Turkish invasion). Late in 1980 tensions with Turkey were somewhat reduced and Greece secured its return to NATO military structure. In addition, the tripartite American-British-Canadian plan of 1978 won Greece's approval as a framework for negotiations, although it was rejected by Turkey as well as by Greek Cypriots.[13]

When the Socialist government of Andreas Papandreou was elected to office in 1981, the foreign policy of Greece shifted. Less inclined to demonstrate Greece's loyalty to NATO and other Western institutions, Papandreou sought to 'internationalise' the Cyprus settlement effort, and adopted a more confrontational approach to bilateral differences with Turkey. This led to a new, and sometimes uneasy, division of labour between Greece and Cyprus, with the latter engaged in intercommunal talks and the former raising the Turkish troops issue in NATO and other international fora. Cyprus, thus was relinked to bilateral Greek–Turkish problems, insofar as the Greek Socialist government insisted that relations between the two NATO allies and the improvement of EC-Turkey Association Agreement could not improve until the Cyprus problem was solved and Turkish troops withdrawn from the island. The critics of this approach, both in and out of Cyprus, charged that it opened the way to and provided the justification for the Turkish Cypriot UDI in November 1983.[14] Interestingly enough, Papandreou's assertive stance was temporarily abandoned in early 1988, when he and Turkish prime minister Ozal conducted talks known as the Davos process, aimed at improving ties through, among other means, confidence-building measures.

In April 1990 the right-centrist New Democracy party returned to power and the new prime minister, Mitsotakis, pledged to renew Greece's efforts to solve the Cyprus problem. It remained Greek doctrine, however, that any attempt on the part of Turkey to extend its control beyond its current lines of occupation would certainly provoke an all-out war. Papandreou had also declared in 1987[15] – and reiterated in 1993 – that in case of a new attack against Greek-Cypriot positions, this would be considered as 'causus belli'.[16]

Militarily, in early 1994, Greece and Cyprus pronounced a joint defence doctrine which was designed to serve as deterrent against further Turkish military action in Cyprus.

Diplomatically, Greece successfully supported Cyprus in its relations with the EU and more particularly made serious efforts to promote Cyprus' application for full membership of the EU:

> The Greek government (in close collaboration with the Government of Cyprus) set two immmediate goals. The first was to secure a firm political commitment as a full member state on the part of the European Union for the accession of Cyprus and try and dissociate the accession process from the solution of the Cyprus problem. The second goal was to agree on a precise date for the commencement of the accession negotiations.[17]

## The US mediation strategy

The Greek Cypriots and the Turkish Cypriots appear to consider the United States as both the cause of much of their misery as well as the only possible external saviour.[18] Relations with the United States after the 1974 crisis were shaped by a conviction in Cyprus that the United States had been too close to the Greek junta, that it could have prevented its coup against Makarios, that it had supported or acquiesced in the Turkish invasion, and that it has paid insufficient attention to solving the Cyprus problem. Relations between Cyprus and the United States were affected dramatically by the 1974 assassination of the United States Ambassador Roger Davies shortly after the Turkish invasion in Nicosia. Yet, pressed by the United States Congress and the aroused Greek- American community, the Nixon and Ford administrations became involved in refugee resettlement and the intercommunal talks.

As the Turkish occupation was consolidated, leading to a long-term division of the island, the Greek Cypriot government continued to have misgivings about the strategic intentions of United States policy. The Cypriots, nonetheless, occasionally pressed for new American initiatives, although none was offered after the 1978 plan. A more activist American policy was institutionalised through the establishment in 1981 of a Special Coordinator in the Department of State. US diplomats have been able to play an active role, albeit behind the scenes, in the UN-sponsored settlement efforts underway since late 1994. 'One explanation may be a mellowing of Cypriot attitudes towards the United States, and a greater appreciation of the constructive role which American diplomats can play, albeit discreetly, in support of the UN talks.'[19] However, efforts by these American diplomats to stimulate discussion about confidence building measures and other intercommunal cooperative projects met with resistance from either the Turkish-Cypriot leadership or the Cypriot government.

Cyprus looked also to the United States Congress and the Greek-American community to correct what it considered a pro-Turkish bias in United States policy.[20] The partial arms embargo against Turkey (1975–8) had injected the Cyprus question in US domestic politics becoming simultaneously 'a serious impediment in the implementation of US strategic plans for the Middle East/ Persian Gulf region.[21]

The new post-Cold War era did not much diminish Turkey's strategic value for the United States. In fact Ankara quickly began to view itself as a regional power with assigned duties beyond its national borders well beyond the 'parochial' Greek–Turkish differences. Today, for the State Department the Cyprus problem seems to be a good demonstration of how maximalist parties can lose good opportunities, which eventually become inoperative. Turkey on the one hand bears the brunt of American criticism for perpetrating 'faits accomplis', which are deemed nearly always irreversible. On the other hand, it steadfastly argues that a constitutional versus territorial trade-off could be an acceptable solution.[22]

Nevertheless the United States continues to be considered as one of the major outside players in the Cyprus dispute. The United States is the only power which can act as a 'catalyst' by bringing the two parties together. But this substantial US ability of decisive intervention has failed to move the problem forward given that procedural and substantive differences between the parties continue to emerge.[23] The 'incapacity' of the United States to promote new initiatives might be due to the classification of Cyprus as a low-to-moderate in importance issue in the overall scheme of American foreign policy, and second, to the strategic importance of Turkey for American interests in the region. Finally, there is the perception that 'constituencies and interest groups' that try to influence US policy on Cyprus represent a stable variable in the US decision-making process.[24]

## The development of a credible deterrent strategy

The Turkish occupation forces amount to well over 30,000 troops without taking into account the 4,000 men of the Turkish-Cypriot forces.[25] In other words, the occupied territories have been transformed into a huge depot of armaments exceeding by far any community defence requirements.[26] Their only military assignment is to safeguard the partition created by Turkish operations. According to the Foreign Affairs Committee of the House of Commons, the continuing military occupation of northern Cyprus constitutes a factor precluding recognition.[27]

On the other side the National Guard of the Republic of Cyprus has 10,000 personnel in active service. Their equipment includes tanks of French and Russian origin, armoured personnel carriers and a wide variety of artillery, air defence missiles and guns, and anti-tank weaponry. According to UN estimates, confirmed by British sources, there remain in Cyprus less than 2,500 Greek troops – mainly officers and non-commissioned officers – in support of the Cypriot National Guard.[28]

There is no doubt that the balance of power still favours the Turkish occupation forces which outnumber the National Guard by 3.5:1 in terms of manpower and 5:1 in terms of main battle tanks. The occupation forces continue to hold complete superiority in the air and the sea.[29]

The Cyprus government has decided to deal with this military inferiority by increasing the credibility of its defence. Two main policies have been adopted: the first seeks to implement an extensive armament programme. Since 1988, successive Cypriot governments have been implementing a programme of arms acquisition to enhance significantly the equipment operated by the National Guard, which has resulted in a large increase in defence spending by the Cypriot government. On 4 January 1997, the Cypriot government decided to sign an agreement with Russia (US $500 million) for the acquisition of some six batteries of advanced S-300 anti-aircraft and anti- ballistic ground-to-air missiles.

The Greek Cypriot annual defence spending amounts to around 7 per cent of the GNP, which is one the highest defence spending rate in Europe with the exception of Croatia and Yugoslavia. Defence expenditure in Cyprus in 1993 reached the highest point ever as a percentage of the GPD, at 7.4 per cent, and in absolute terms. Yet, as Aristotelous notes, 84 per cent of this sum was allocated to payments of loans and other financial obligations related to the purchase of arms and equipment during previous years.[30]

The leader of the Turkish Cypriot community, Rauf Denktash, has accused the government of Nicosia of being engaged in a costly rearmament programme, which, according to

Turkish analysts, raises fears that Nicosia has not abandoned a military option to the Cyprus problem.[31] It should however, be stated that Cypriot defence expenditures are designed to increase the deterrent value that would reduce Turkish temptations to make further inroads on Cypriot territory. The long range goal of applying an effective strategy of deterrence in the Republic of Cyprus leads to efforts by a sovereign and independent state to arm itself for self-defence.

The second policy has been aimed at increasing defence credibility directed towards building up closer military ties with Greece. To this effect Greece and Cyprus decided in 1991 to set up a joint military committee, create a special command responsible for the Cyprus' theatre in the Greek Minsitry of Defence, strengthen cooperation and communication, and enhance military planning by providing the National Guard with air and naval support.

In an effort to favour defence over offence and dampen adventurous initiatives in Cyprus, Greece and Cyprus have recently adopted a new defence doctrine which incorporates Cyprus into an extended perimeter of the Greek defence.[32]

In 1994, after drawn-out discussions among experts, academics and the military, Athens agreed with Nicosia to enhance stability and maintain the deterrent capability of the Republic of Cyprus.[33] The details involving coordination in both the Greek and Cypriot defence perimeters as well as the operational responsibilities of the respective armed forces begun being specified in January 1994.

According to the Greek point of view the risks of an incident in Greek–Turkish relations are so high and the vulnerability of the Cypriot defence so evident that something had to be done in the direction of counterbalancing present military asymmetries. Thus, in the future it is expected that the defence of Cyprus and Greece will remain in a forward theatre of war that will be much closer to Turkey than to the Greek mainland. The credo of the Greek Cypriot common defence strategy, which is expected to be operational as of 1997, is twofold and aims at:

a) granting military protection to the Cypriot Republic and
b) facilitating a diplomatic solution of the crisis by rendering extremely costly any calculated rewards for the aggressor.[34]

## THE PROSPECT OF ADMISSION TO THE EUROPEAN UNION

On 3 July 1990, the government of Cyprus submitted to the Council of the EC an application for membership in an effort to accelerate closer relations with the EC. Cyprus had been linked with the EC through an Association Agreement signed in December 1972. The Agreement came into force on 1 June 1973 and was to be accomplished in two stages, the first of which was to expire in 1977. This was to be followed by a second stage, with the final objective of a customs union to be achieved by the beginning of the 1980s. However instead of a second stage, there had been several successive temporary protocols which had extended stage one with only minor improvements.[35]

Cyprus' relations with Brussels have been troubled by the uncertainty of the political situation on the island and the EC's preference to avoid entanglement in political disputes. Throughout the years of the division of the island EC policy was to deal with the Republic government as the legal authority, but at the same time to state that the benefits of the Association Agreement must extend to the entire island and its population.

The acceptance by the EC that a political solution to the Cyprus problem should no longer be a prerequisite of further development of EC/Cyprus relations, together with Greek insistence, and the abandonment by France, Italy and Spain of their previous reservations, paved the way for the conclusion of the negotiations for the Customs Union Agreement signed on 22 May 1987. It was to be implemented in a period of fifteen years. The Agreement created a climate of optimism in Cyprus. The majority of the political parties considered the Customs Union Agreement as a preliminary step to full admission.[36]

On 4 July 1990, the Government of Cyprus submitted its application for membership to the EC. The submission of the application was facilitated by the speedy and impressive recovery of the Cypriot economy, despite the destruction that had been caused by the Turkish 1974 invasion and subsequent occupation.

For the government of Cyprus there are both economic and political reasons for submitting the admission application.[37]

In the economic field, some main advantages are access to a large and dynamic market, which is already Cyprus' main trading partner; participation in the full range of EU financial mechanisms and funds; and an opening of the doors of EU to all Cypriots in all fields. There was also a concern that with the creation of the single market and with the accession of Greece, Spain and Portugal in the 1980s, the EC was becoming self-sufficient in Mediterranean produce and this would have a negative impact on Cyprus exports. Finally an obvious advantage is Cyprus participation in the Common Agricultural Policy. The inclusion is advantageous to Cyprus to the extent that it waives present EU quotas on Cypriot agricultural exports.[38]

These economic concerns and opportunities are augmented by the political benefits that the Greek Cypriots could obtain from full membership in the EU. The application could be seen as a tactical move as well as a strategic one. It would be tactical as a move intended to give new momentum and new incentives to the Turkish side to achieve progress in talks. Cyprus seeking full membership would inevitably involve the EU more closely in the search for a negotiated Cyprus settlement on terms which Greek Cypriots think will be more favourable to them. 'The Cypriot government has gradually come to see the EU as a potentially more 'credible' instrument of conflict management with respect to Greco-Turkish disputes, one that would soon take over the role of NATO'.[39]

From a strategic point of view, Cyprus stands to benefit immensely from EU admission for two main reasons. First, as a member, Cyprus would participate in EU policy formulation, and its views would be heard in EU organs at all levels of hierarchy. An entry into the EU would strengthen the sovereignty, independence and unity of the country by weakening efforts to partition Cyprus. It would discourage all those who at present constitute a real threat to the security and peace of its inhabitants and encourage a climate of coexistence and cooperation. Second, the implementation of the EU regulations would guarantee human rights and in particular the fundamental rights of movement, establishment and property.[40]

In its opinion on Cyprus' application to join the EU at the end of June 1993, the European Commission delivered a positive message to the divided island as an incentive to try to break the Cyprus deadlock. On the one hand it emphasised that accession negotiations could open as soon as there was sufficient certainty about the prospects for unification of the island, on the other hand it added the proviso that if no prospects for agreement were in view by 1 January 1995, it would review the situation and assign responsibility for the future.[41]

In November 1993 substantive talks opened between the government of Cyprus and the Commission and were successfully completed in February 1995. The primary objective of these talks, which covered a broad range of subjects, was to allow the Cypriot authorities to familiarise themselves with all the elements that constitute the acquis communautaires and to help Cyprus harmonise its legislation and policies to those of the EU. Meanwhile in June 1994 the European Council at Corfu concluded that 'an essential stage in Cyprus' preparations for accession could be regarded as completed'. It also decided that 'the next phase of enlargement of the Union will include Cyprus and Malta'. The European Council meeting in Essen on 19 December 1994 confirmed the Corfu conclusions. The Corfu and Essen conclusions are important, not only because they affirmed the eligibility of Cyprus for membership, but also because the issue of Cyprus' integration to the EU has been clearly disassociated from the solution of the Cyprus problem, since the latter is not considered a precondition for its accession to the EU.

On 6 March 1995, when the European Union concluded a customs union agreement with Turkey, it was also agreed that the negotiations for Cyprus accession would start six months after the successful conclusion of the 1996 Intergovernmental Conference. It was hoped that this promise of future membership could work as a catalyst to end the stalemate in negotiations between the Greek and Turkish Cypriots on an overall agreement[42].

The interest of the EU on Cyprus was also manifested with the appointment by the Council on 20 December 1993 of Serge Abou, a Senior Official of the European Commission, as an observer in the intercommunal talks. Abou submitted three reports in which, inter alia, he confirmed that, as it was clearly indicated in the UN Secretary-General's report

(31.5.1994), the lack of progress in the intercommunal talks was due to the lack of political will on the part of the Turkish Cypriots.

In accordance with the 6 March 1995 conclusions, the 16th Association Council meeting between Cyprus and the EU held on 12 June 1995 agreed on the establishment of a structured dialogue between the EU and Cyprus, and on certain elements of the strategy to prepare Cyprus' accession to the EU. The structured dialogue was to include meetings between the heads of state or of government, on the occasion of the European Council as well as meetings and talks at ministerial and other levels on issues falling within the Common Foreign and Security Policy. As a result Cypriot officials participated in numerous meetings with officials from the member states on issues such as global disarmament and arms control, human rights, terrorism, drugs, former Yugoslavia, Eastern Europe, the OSCE etc.

In June 1995, the Fourth Financial Protocol of a total amount of 74 million ECU was signed, covering the period up to 1998. It is interesting to note that the Protocol had a pre-accession character and aimed to promote the development of the Cypriot economy and the objectives of the EEC–Cyprus Association Agreement, to facilitate Cyprus' economic transition with a view to its accession to the EU and to support efforts to promote an overall settlement of the Cyprus problem.

There is no doubt that accession to the EU is the one of the biggest challenges for the Cypriot foreign policy since 1960.[43] But do all these recent positive developments in Cyprus–EU relations really mean that Cyprus is on its way to becoming a full member of the EU? Or do some serious difficulties still remain which might cancel the prospect of accession?

From the economic point of view Cyprus membership is an exception. All too often the accession of new members from the Mediterranean area has involved great costs to the EU. In the case of Cyprus the opposite would be true. With a per capita income of about $13,500 in 1995, Cyprus is comparable to that of many European Union member states. The high standard of living is coupled with other socioeconomic indicators, such as high life expectancy and the high enrolment ratios in all the educational levels. The economic indicators show that Cyprus is not only prepared to join the EU but is also in a position to proceed to the third stage of the Economic and Monetary Union (EMU) meeting most of the economic criteria that EU member states must meet before they can join the EMU. The rate of growth in 1996 has been around 2.5 per cent and in 1997 it is expected to rise to 3.5–4per cent. The inflation rate was 2.5 per cent in 1995 and 3 per cent in 1996, mainly due to expected higher agricultural prices in comparison with the usually low levels of 1995. Unemployment was at 3 per cent. The Government budget deficit stood at 2.9 per cent of the Gross Domestic Product. The public debt was at 54 per cent of the Gross Domestic Product and the Cyprus pound had shown remarkable stability and had been linked since 1992 without any problems, to the narrow band (plus or minus 2.25 per cent) of the Exchange Rate Mechanism (ERM). The economy of Cyprus is geared towards light industry and services – mainly international tourism, though comparative advantages exist in agriculture, manufacturing and financial services, particularly in offshore banking. Industry accounts for three-quarters of exports of goods. The large trade deficit is more or less compensated for by the earnings from tourism. More than half of its foreign trade, both imports and exports, is conducted with the EU.

Given this Cyprus economic performance, the European Commission identified in 1993 that the 'adoption by Cyprus of the Union's '*acquis communautaires*' will pose no insurmountable problems. The Commission is not underestimating the problems that the economic transition poses. However, the economy of the island has demonstrated an ability to adapt and seems ready to face the challenge of integration, provided that the work already started on reforms and on opening up to the outside world is maintained, notably in the context of the customs union.'

It is estimated that the accession negotiations between Cyprus and the EU will not last more than two years because Cyprus will be able to harmonise its legislation without difficulties. Problems, of course, may arise during the negotiations particularly in the fields of the status of the offshore companies, banking, shipping and agriculture.

From a political view, the main problem of accession is the continued Turkish occupation of Northern Cyprus. Many member states continued to consider that Cyprus admission to the EU would make a settlement of the Cyprus dispute more difficult. The British government, for example, believed that the prospect of EU accession should act as a catalyst for a settlement in the months leading up to the opening of negotiations with the EU. 'If there were no settlement, EU accession for Cyprus would be immeasurably more difficult, making negotiations a complicated and long drawn out process.[44] The real fear, however, was that the impact of Cypriot accession might well be the final straw that would drive the Turks to renewed open conflict with Greece and a curtailment of Turkey's pro-European line.[45]

The Cyprus government, as well as Greece, argue that the accession negotiations should not be linked with the solution of the Cyprus problem. That does not mean that they are not eager to enter into a fair, functional and mutually acceptable settlement at the earliest possible moment. The prospect of Cyprus' membership is probably the only strong and persuasive pressure which has been exerted upon Turkey with regard to Cyprus since 1974. This would not serve its purpose however if the solution to the problem is considered a prerequisite for accession. It will only encourage those intransigent elements in Turkey, and will make Cyprus a hostage of Turkish mainland politics. Cyprus should not be punished because it has been the victim of Turkish aggression and military occupation.[46] Therefore, if the Cyprus problem has not been settled in the next few years, Cyprus should become a full member of the EU and just as the EU–Cyprus Customs Union applies to all of Cyprus, but in practice is implemented in the southern part, accession will apply to the whole of Cyprus but will be implemented immediately in the southern part.[47]

## CONCLUSIONS

None of the foreign policy stategies which Cyprus has adopted over the last twenty-two years have produced positive results for finding a just solution to the Cyprus problem.

However, some could be considered as partially successful. For example, the Cypriot strategy of committing the EU to open negotiations in the near future for its admission to the European Union has been proven successful due to the assistance provided by the Greek government. The EU, despite the fact that some of its member states were unwilling to consider seriously the Cyprus application for membership, is now politically committed to open negotiations with Cyprus within six months of the completion of the Intergovernmental Conference.

On the other hand, the strategy of encouraging US mediative efforts in the Cyprus dispute has not proven very fruitful to date. The hopes that arose from the US declarations of a new world order after the end of the Cold War in Europe and of the second Gulf War never materialised. It seems that for US foreign policy the Cyprus problem is not among its highest priorities in the area. Washington is more interested in supporting the preservation of a pro-western regime in Turkey and of encouraging a close relationship of Turkey with the West rather in exercising its political influence on Ankara to make concessions on Cyprus. That does not mean that in the near future Washington will neglect the Cyprus question. On the contrary the US appears to believe seriously that a rapprochement between Greece and Turkey is only possible if and when substantial progress in Cyprus is achieved.

Under these circumstances, and given the fact that all the UN initiatives for the resumption of the intercommunal negotiations have failed for the time being, either because of the Turkish Cypriot instransigence or because of the Greek Cypriot unwillingness to accept an unworkable set of ideas proposed by the UN Secretary-General, the Cypriot government began to consider alternative policies. One of them was to link Cypriot defence requirements with the Greek defence strategy. Although strictly speaking militarily the new doctrine of 'extended deterrence' of Greece and Cyprus is not going to change the balance of power on the island for the foreseeable future, it has however a positive psychological and political effect for the Cyprus population. It has created an atmosphere of cordial relationship between Athens and Nicosia which was not always the case during the past. Of course the new doctrine alone is not enough to deter Turkey's military superiority.

Cyprus hopes for a peaceful solution of the problem depend in the short term on the mediation efforts pursued by the US, and in the long term on full membership in the EU.

## NOTES AND REFERENCES

1. The Treaty confers to Greece, Turkey and Great Britain the status of guarantor powers with the specific obligation to ensure Cypriot independence, territorial integrity and security (art.II.a).
2. M. Michael, 'The road to Vienna: Intercommunal talks 1974–1977', *Cyprus Review*, Vol. 4, no. 2, 1992, p. 115.
3. For the intercommunal talks during this period see P. Polyviou, *Cyprus: Conflict and Negotiation 1960–1980*, London, Duckworth, 1980.
4. An essential part of the confidence-building measures is the rehabilitation of the Varosha area north of UN-administered buffer zone, at present surrounded by territory controlled by the Turkish Cypriot administration, as an area for trade and intercommunal contacts and the re-opening of Nicosia international airport to civilian and cargo traffic. See Report of the General-Secretariat. UN S/26026 1.7.1993. For the failure of the confidence-building measures see H. Brey, 'A solution to the Cyprus Question: Options and obstacles', *RFE/RL Research Report*, vol. 3, no. 28, 15 July 1994, pp. 23–5.
5. M. Michael, op.cit., p. 115.
6. Van Coufoudakis, 'Cyprus: Domestic Politics and Recent UN Proposals' in C. Ioannides (ed.), *Cyprus Domestic Dynamics, External Constraints*, New Rochelle, A. Caratzas, 1992, p. 15.
7. Van Coufoudakis, 'United States and the UN since 1960', in *Etudes Helleniques*, op.cit., p. 44.
8. I. Nikolaou, 'La Grèce en Quête d'une Défense Flexible', in P. Buffotot (ed.), *La Défense en Europe – De la Guerre du Golfe au Conflit Yougoslave*, Paris, La Documentation Française, 1995, pp. 93–102.
9. V. Coufoudakis, 'Greece, Turkey, Cyprus and the United States in the Changing International Order', in H. Psomiades

and S. Thomadakis (eds), *Greece, the New Europe, and the Changing International Order*, New York, Pella, 1993, p. 395.

10. Resolution 3212 (XXIX) of the General Assembly (1, November 1974), and Resolutions 365 of the Security Council (13, December 1974) and 541 (18, November 1983) call for the effective and urgent implementation of its resolutions on Cyprus.
11. J. Joseph, 'The International Power Broker: A Critical View of the Foreign Policy of Archbishop Makarios', *Mediterranean Quarterly*, vol. 2, no. 2, 1992, p. 31.
12. V. Coufoudakis, 'Cyprus, The Commonwealth and the UN since 1960', *The Cyprus Yearbook*, 1993, p. 18.
13. For the American-British-Canadian plan see S. Mayes, 'Can the Cyprus problem be solved?', *Round Table*, No. 273, 1979, pp. 81–7.
14. D. Constas, 'Greek Foreign Policy Objectives: 1974–1986' in S. Vrionis (ed.) *Greece on the Road to Democracy: From the Junta to PASOK 1974–1986*, New Rochelle, A. Caratzas, 1991, p. 45.
15. According to the Prime Minister's declaration '... should Cyprus perish, Greece will eventually follow'. See the text of Parliamentary Debates, 23, January 1987, p. 2915.
16. *Phileleptheros*, 20, January 1994.
17. Y. Kranidiotis, *The Enlargement of the European Union and the Mediterranean: The Challenge for Cyprus*. The 1995 Canon Newham Memorial Lecture 18 December 1995. Nicosia: The English School Old Boys and Girls Association, 1995, p. 6.
18. E. Laipson, 'The United States and Cyprus: Past Policies, Current Concerns', in N. Salem (ed.), *Cyprus. A Regional Conflict and its Resolution*, London: Macmillan, 1992, p. 90.
19. E. Laipson, 'Cyprus: A Quarter Century of US Policy' in J. Koumoulides (ed.), *Cyprus in Transition 1960–1985*, London: Trigraph, 1986, p. 78.
20. C. Hackett, 'The Role of Congress and Greek-American Relations' in T. Couloumbis and J.Iatrides (eds), *Greek-American Relations*, New York, Pella, 1980, pp. 131–147.
21. M. Evriviades, 'The US and Cyprus: The Politics of Manipulation in the 1985 UN-Cyprus high level meeting', *Cyprus Yearbook*, 1992, p. 17.
22. Van Coufoudakis, 'United States and the UN since 1960' in *Etudes Helleniques*, op.cit., p. 46.
23. C. McCaskill, 'The United States and Cyprus from 1974 to 1991' in C. Ioannides (ed.), *Cyprus. Domestic Dynamics, External Constraints*, New York: A. Caratzas, 1992, p. 123.

24. E. Laipson, 'The United States and Cyprus: Past Policies, Current Concerns', op.cit., p. 93.
25. See *Security Issues of Cyprus. Part 1. Seminars and Roundtables*, No. 6, 1989, Hellenic Foundation for Defence and Foreign Policy (in Greek).
26. The Turkish occupation forces amount to more than 30,000 soldiers allocated in two divisions detached from the 3rd Turkish Army Corps based on Erzincan, Nicosia: KYKEM, 1991, p. 37.
27. House of Commons, Foreign Affairs Committee, Report on Cyprus, 7 May 1987.
28. *Military Balance 1983–84,* London: IISS- Brassey's, pp. 77–78.
29. See A. Aristotelous, *Ellada, Tourkia, Kypros – Stratiotika stichia kai arithmi 1990/01 [*Greece, Turkey, Cyprus – Military data and numbers 1990/91], op.cit.
30. A. Aristotelous, 'Defence expenditure and the military situation in Cyprus', *Cyprus Yearbook*, 1993, p. 39.
31. Z. Necatigil, *The Cyprus Question and the Turkish Position in International Law,* Oxford: Oxford University Press, 1993, p. 392.
32. K. A. Pisiotis, 'L'OTAN et la persistance du conflit greco-turc', *Politique Etrangère*, winter 1993/94, p. 914.
33. *Phileleptheros*, 23.1.1994.
34. For the new doctrine of the single defence area between Greece and Cyprus see A. Aristotelous, *The single defence area: Greece–Cyprus*, Nicosia: Cyprus Centre of Defence Studies, 1995 (in Greek) and T. Dokos, 'The doctrine of the single defense area', *Institute of International Relations Yearbook*, 1996, pp. 143–156 (in Greek). For a critical view see K. Zepos, 'Greek-Turkish relations and the single defence area', *Institute of International Relations Yearbook*, 1996, pp. 126–134 (in Greek).
35. C. Tsardanidis, 'The EC–Cyprus Association Agreement: Ten Years of a Troubled Relationship, 1973–1983' *Journal of Common Market Studies*, Vol. 22, No. 4, June 1994, pp. 351–376.
36. C. Tsardanidis, *The Politics of the EEC–Cyprus Association Agreement 1972–1987*, Nicosia: Social Research Centre, 1988; C. Stefanou, C. Tsardanidis, 'The EC factor in the Greece–Turkey–Cyprus Triangle', in D. Constas (ed.), *The Greek-Turkish Conflict in the 1990s. Domestic and External Influences* London: Macmillan, 1991, pp. 207–230.
37. J. Redmond, *The Next Mediterranean Enlargement of the European Community: Turkey, Cyprus and Malta*, Aldershot, Dartmouth, 1993, pp. 62–97.

38. Of course there is no doubt that Cyprus will confront difficulties in adapting to the full membership of the EU. See P. Nicolaides, 'Cyprus and the European Community: Looking beyond 1992', *Cyprus Review*, vol. 2, no. 1, 1990, pp. 44–80, and H.-J. Axt, 'Zypern-ein Beitrittskandidat der Europaischen Union implikationen fur die Insel, die Region und die Union', *Sudost-Europa*, vol. 44, no. 5, 1995, pp. 259–279.
39. D. Theophylactou, *Security, Identity, Nation Building*, Aldershot, Avebury, 1995, p. 117.
40. *Cyprus and the European Community*, Nicosia Euraconsult, 1991.
41. S. Calleya, 'Prospects for EC Expansion in the Mediterranean Region', *Southern European Yearbook*, 1993, p. 42.
42. Turkey and the Turkish Cypriot leadership have reacted angrily to the prospect of admission of Cyprus to the EU. Denktash has stated that he was not consulted – and therefore that the equal status of the Turkish Cypriots has been ignored. The Turkish Foreign Minister, M. Karayalcin, stated that if accession negotiations were held only with the Greek-Cypriot side of the island, Turkey 'would be left with no option other than to take steps towards achieving a similar integration with the "Turkish Republic of Northern Cyprus", *Financial Times*, 8.3.1995. Tansu Ciller, then Turkey's Prime Minister, visiting the United States in April 1995, suggested that a solution to the Cyprus problem would be found after Turkey becomes a member of the European Union, *International Herald Tribune*, 19.4.1996.
43. J. Joseph, 'Cyprus at the Threshold of the European Union', *Mediterranean Quarterly*, vol. 7, no. 2, 1996, p. 116.
44. S. Wright, *Cyprus' Accession to the European Union – A British View*. Paper presented at the Conference on Cyprus in the European Union organised by the European Institute of the London School of Economics and Political Science and the Hellenic Centre, 30–31 October 1996, p. 3.
45. J. Redmond, 'Secondary Implications of the Accession of Cyprus to the European Union', *International Spectator*, vol. 30, no. 3, 995, p. 35.
46. Y. Kranidiotis, *The Accession of Cyprus to the EU and the Search for a Peaceful, Just and Lasting Solution to the Cyprus Problem*'. Paper presented at the conference on 'Cyprus in the European Union'.
47. Y. Kranidiotis, *The Enlargement of the European Union and the Mediterranean: The Challenge for Cyprus*, op.cit., (note 17).

# 8 Malta's Foreign Policy in the 1990s

Roderick Pace

## INTRODUCTION

The theoretical tools for the analysis of Malta's foreign policy draw from two basic and complementary trends in the study of international relations: the theories of international relations (IR) and foreign policy analysis (FPA). Reference will be made to the evolution of Malta's foreign policy since independence with an eye to showing the roots of some of today's policy decisions and choices. The next step will involve extrapolating from these findings to determine the likelihood of these trends being maintained, or changed, to the end of the 1990s. The task will be encumbered by the fact that in the case of Malta, one is dealing with the behaviour of a micro-state in the international system and the theories of the behaviour of micro-states in international relations remains rather undeveloped. There appears enough circumstantial evidence to suggest that this can be blamed on the influence of realist theory in international relations. Once power in its most general meaning becomes the central concept of analysis, then micro-states, which individually and collectively have so little power in the world, tend to pale into insignificance as a subject of academic analysis. The traditional paradigm, the 'billiard ball' model, makes little distinction between states, since this model projects them as possessing identical properties. However we must not loose sight of the fact that realists and neo-realists acknowledged the crucial importance of the variation of power between states. Morgenthau pointed out that national power derives from a number of variables which account for the difference in the power of states in the international system.[1] He further adds that power is relative (has to be measured in relation to the power of other states in the system), is not permanent and cannot be solely linked to one factor but has to be tied to other power factors of the same nation.[2] Waltz refers to the inequality of states as being inherent in the states system and ineradicable. He credits this inequality as being conducive to the preservation of international peace and stability.[3]

The analysis of the foreign policy of Malta in the 1990s is also hampered by two other considerations which are not unique to the analysis of micro-state behaviour, namely the level of analysis problem and the sources of policy analysis. For the sake of brevity a comprehensive analysis of Malta's foreign policy from the various standpoints of foreign policy analysis – such as the bureaucratic, psychological or systems approaches to mention a few–will not be attempted here. Some allusions to further research and application of these methods will be made in this text. One important question is whether the tools applied to the analysis of the foreign policies of developed states have any relevance when applied to small mostly developing states. The answer has been partially answered by Hill,[4] who pointed out that one can treat different types of states within the same set of concepts, with of course different emphasis. Therefore while the bureaucratic analysis approach may be more important when dealing with advanced states with full-fledged bureaucracies, it is not an element to be completely discarded in the case of developing states. The other problem, the one concerning sources, arises from the fact that most of the analysis in this chapter is based on public documents and little access has been gained to the unpublished sources and documentation which throw more light on how decisions have been arrived at. Indeed, throughout the development of International Relations as a discipline, this source of information has been at once both the most precious and the most elusive for scholars. Furthermore, the operational environment of the foreign policy decision-makers assumed below might be more of a conjectural one, as viewed from outside. This contribution will no doubt rationalise the whole foreign policy process in Malta and present it in a more coherent form than was perhaps the case in reality, given that politicians are very often constrained to take political decisions in reaction to stimuli both from their domestic as well as international environment.

The undertones of the present analysis will however be those of the rational actor model. The overlap between the two environments, the domestic and the external, are crucial so that in fact we are dealing with one and not two, though for analytical purposes it pays to classify phenomena as originating in the domestic or in the international environment. The linkage between the 'two environments' has for long now been accepted in the study of international relations. As Rosenau showed in his elucidation of 'linkage politics', there is an intimate relation between national and international politics: 'Where the functioning of any political unit was once sustained by structures within its boundaries, now the roots of its political life can be traced to remote corners of the globe.'[5] Later, he explored the notions of 'across-systems analysis', while further elaborating the efficacy of linkage politics.[6] As to the question of elite images, so essential to the understanding of some foreign policy decisions, the analysis will be based on public declarations made by Maltese elites. Notwithstanding this however, there is always the problem that such elites may operate two separate, at times conflicting, agendas: the hidden one, in the direction of which they are actually moving, and the public agenda, which is the one employed to garner public support for otherwise unpopular policies, or to camouflage the hidden agenda.

## SMALL STATE THEORY IN INTERNATIONAL RELATIONS

To begin with a discussion of Malta's foreign policy merits reference to 'small'or 'weak' state theory. This approach can be useful in determining which of Malta's biases in its actions in the international arena are linked to the peculiarities of smallness, and hence tend to be more resilient in the long run, and which are contingent, varying in reaction to short-run changes or to ideological differences and choices of the competing elites responsible for Malta's foreign policy. The notion that there is a link between nation-types and foreign policy behaviour was first put forward by Rosenau[7] in 1966. Empirical evidence has not suggested that Rosenau's eight nation types can be used to explain foreign policy behaviour. However, a number of studies suggest that physical size of a state best accounts for foreign policy behaviour.[8] The first problem that small state theory has to confront is whether there is sufficient reasons to distinguish 'small state theory' as such from the general theories explaining the behaviour of states in general in the international system. Obviously differences in power, territory, population, military strength, human resources, technological potential, cultural and other variables, between small and large states undermines the conventional wisdom that small states are nothing but large ones writ small. Hence a specific small state theory which is nevertheless intellectually linked to the general theories of international relations is perfectly in order. The approach which will be used in this analysis will be to employ the tenets of this theory as it presently stands to Malta's foreign policy. Logically it will be fallacious to move in the opposite direction of using Malta's case to construct theory since this would entail moving from the particular to the universal.

The next problem to be confronted is certainly one of definition. Reference has been made in the literature to 'small states', 'smaller states', 'micro-states' and 'small powers', 'dependent' or 'weak states'. The descriptive term 'Power' itself has been further graduated into 'small', 'medium', 'large' and 'super'. Vital,[9] quoting Aron, rightly points out to the use of 'small power' as a contradiction in terms. Handel[10] looked on power both as a continuum as well as a hierarchy, metaphorically a pyramid with the superpowers at the top and lesser states at various levels below, with the weakest, that is the mini-states, at the foot.[11] He developed a theory of weak states to mean small states, though 'weak' is very often used equivocally in his text to cover states which may not be small, but are nevertheless weak. Handel points out to the difficulty of defining groups of states in the international hierarchy using one concise, precise and elegant statement.[12] Waschkuhn confirms: 'a convincing definition of small states and especially micro-states has not been found, neither by scientific nor by political means.'[13] Attempts to classify small states according to population size[14] have been deemed inadequate because they exclude other variables, such as economic and military strength, territorial size and geostrategic position, which either singly or in combination may be an important source of influence in world politics. Hence a lot of weeding out and ordering in the theoretical concepts so far elucidated in the literature is still needed. For example, a small state might be weak in terms of territory, population and military capabilities but strong economically and in terms of the influence it exercises. Conversely, a state may be strong in terms of the three variables just mentioned but economically weak and in a state of dependence. States in fact can be classified according to any number of permutations of variables than the ones indicated here.

The treatment of smaller polities or states has varied in the literature while references to them have generally tended to be sporadic until well into this century. The extent of the literature dealing with small states is limited by any standards and can be divided into three distinguishable genres: that which idealises small states; that which regards them as extremely vulnerable; and that which looks upon them as possessing both strengths and weaknesses. The emphasis has also varied from political/strategic to economic.

From the standpoint of politics and public policy analysis, the most central problem is that in world politics the small state is handicapped by the lack of personnel to man its policies as well as by a low level of information sources to formulate them, while being constrained to participate fully in world diplomacy. Doubts have been cast in the past as to whether small states, especially micro-states, can indeed handle their participation in world organisations such as the United Nations.[15] More seriously, small states are more vulnerable to threats to their security from stronger neighbours. This becomes even more true if this vulnerability is compounded by economic weakness (inviting internal meddling and external manipulation) or an essential geo- strategic position.[16] The small or weak state may seek security via alliances, cooperation, coordinated policies and support from among smaller states. However these tend to be ineffectual in deterring serious challenges.

To sum up the characteristics of weak (small) states, reference can be made to Handel's construction of two ideal types of state, the 'almighty' superpower on the one hand and the weak state on the other hand.[17] According to this scheme a weak state is one which is small in terms of population and area. Economically it has a small GNP in absolute terms, little or no heavy industry, a high degree of specialisation in a narrow range of products, a small domestic market and therefore a high degree of external dependence in trade, high dependence on foreign capital and very low R&D in absolute terms. Militarily, the small state cannot defend itself against external threats and has a high or total dependence on external help, has a total or very high weapon acquisition in foreign countries, a high proportion of its military strength always mobilised and a very low longer range war potential. In the international system, the small state has a limited scope of interest, normally confined to neighbouring and regional areas, little or no influence on the balance of power, is mainly passive and reactive in foreign policy, minimises risks especially vis à vis the powers, can be penetrated relatively easy and exhibits strong support for international law, norms and international organisations.

## FOREIGN POLICY ANALYSIS APPROACH

One last theoretical hurdle that needs to be overcome before the main analysis is undertaken is the question of foreign policy analysis tools which will be applied in this chapter. Hey[18] suggested that case studies are a promising avenue for expanding understanding of dependent foreign policy and elaborates four main models of a dependent state's foreign policy: consensus, compliance, counterdependence and compensation. Thus, *consensus* between core and periphery elites leads to foreign policy alignment; *compliance* means that the dependent states would comply with core wishes in exchange for economic rewards; *counterdependence* means that a dependent state's anti-core foreign policy behaviour is a hostile reaction to a state's dependence; *compensation* refers to the anti-core foreign policy stance which a state may take to maintain internal legitimacy.

A concise overview of the development of Foreign Policy Analysis (FPA) in the literature to the post Cold-War setting is the one provided by Hudson and Vore.[19] The insights provided in this article are most instructive. Of relevance to the limited scope of this chapter will be the use of foreign policy decision-making analysis and foreign policy context, not to mention the attempt to bridge foreign policy analysis with the theories of international relations, reference to which is also made by Hudson and Vore,[20] in this case to link small state theory with FPA. Relevant too are the 'distinctive theoretical commitments' some of which are: to look below the state level of analysis, to build middle-range theory at the interface between actor-general theory and the complexity of the real world, the pursuit of multicausal explanations spanning multiple levels of analysis, the utilisation of theories and findings from across the social sciences, and viewing the process of foreign policy decision-making as a subject of equal importance to the output.[21] More pointedly, the analysis will centre on issues raised by East in his article 'Size and Foreign Policy Behaviour'.[22] Accordingly, East abstracted from the works of various authors (Morgenthau, Aron, Organski and Liska) the following foreign policy behaviour patterns associated with small states:

- Low levels of overall participation in world affairs;
- high levels of activity in intergovernmental organisations;
- high level of support for international legal norms;
- avoidance of the use of force as a technique of statecraft;
- avoidance of behaviour and policies which tend to alienate the more powerful states in the system;
- a narrow functional and geographic range of concern in foreign policy activities;
- frequent utilisation of moral and normative positions on international issues.

He added that these patterns of small state behaviour correspond to what may emerge from a rational or 'conventional' model which holds true of any state with limited resources and a limited international potential, i.e. small states exhibit low-profile courses of action, seeking to minimise risks and their expenditure of scarce resources (in terms of manpower, military capabilities and hard currency). It is also assumed that foreign policy actions of small states are the results of decisions arrived at by a decisional unit, with an adequate information base and a long-range policy perspective.

However, one may begin the analysis by making the opposite assumptions, i.e. use an alternative model, namely that there is a fundamental difference between the foreign policies of small and large states based on the following criteria:

(i) small states have a smaller proportion of an already small resource base to devote to foreign policy and thus they seek actions which are more economic – acting in IGOs, international conferences, regional organisations, multilateral diplomacy, etc.;
(ii) as a consequence of their limited resource base fewer persons are engaged in monitoring international events and executing foreign policy decisions;
(iii) small states will be less active and differentially active in various areas of policy;
(iv) small states would be slower in perceiving events and developments in the international system, i.e. would be unlikely to perceive early warning signals indicating new developments;
(v) as a consequence of the previous point, a small state does not enjoy the luxury of engaging in early, low-level, ambiguous behaviour when trying to take effective action to influence a situation, because by the time it has perceived the signals the situation would have been reached where it has to take definite, unambiguous and high-risk action. This alternative model leads to one fundamental conclusion which differs from the 'conventional' one, namely that small states are more prone to high-risk behaviour.[23]

## MALTA'S FOREIGN POLICY

East's conclusions combined with the small state theory outlined above present an intriguing point of departure for the analysis of Malta's foreign policy in the 1990s. In other words how does Malta's behaviour in the international arena fit the theoretical models discussed above? Further, if its behaviour differs, what are the general theoretical implications? Malta commenced the decade a neutral and non-aligned state, with special ties with two of its more powerful neighbours namely Italy and Libya, and with a determined policy to join the European Union. Following the change in government in

October 1996, which saw the return of the Labour Party to power after an absence of nine years, Malta has switched its policy on EU membership, declaring its interest in concluding a free trade area agreement with the Union, as well as a Security Guarantee protocol. The Maltese Prime Minister did not attend the Dublin European Council meeting of 14 December 1996 with the other EU applicants. However, at the same time Malta has not retracted its membership application.

Malta's foreign policy in the nineties can be better understood in the context of the policy decisions taken since independence in 1964 which can be summed up as follows:

(a) the realisation that since it is small it will always be a 'consumer' of security, not a producer, relying very much on external aid to maintain its defence. Hence its foreign policy exhibits a perpetual quest for economic and political security.
(b) maintaining good neighbourly relations;
(c) attempting to influence the external environment to make it more congenial to Malta's specific needs by strengthening international organisations, norms and agreements.
(d) Attempting to achieve all these aims in the most 'economical' way by targeting action towards groups of countries and by generally, though not always, seeking to achieve aims in concert with others.

## The Quest for Economic and Political Security

Since independence Malta's overriding concern has been the consolidation of security in its wider meaning,[24] the strengthening of independence and the national identity through the lessening of dependence especially on a single external power. The means to achieve this security have differed markedly from time to time in accordance with the differing ideologies of the political elites. Thus in the period 1964–71, under a Nationalist Government, Malta's foreign policy was pro-Western; in the period 1971–87, under a Socialist government, Malta's foreign policy was officially neutral and non-aligned, yet openly anti-Western; from 1987 onwards, again under a Nationalist administration, there is a return to the pro-Western policies of 1964–71, this time strengthened by a keen interest in joining the European Union. Though it is still early to predict how the new Labour Government elected in 1996 will perform, it appears to have parted significantly with old practices and though it says it does not think Malta should join the European Union in the foreseeable future, it nevertheless wishes to have the closest relations possible with Europe.[25] The main thrusts of Malta's foreign policy to achieve economic viability and security have been: (i) the conclusion of defence and financial pacts followed by the attempts to obtain recognition of its neutrality; (ii) the approach to the European Community; (iii) relations with neighbouring states and (iv) action in international organisations and fora.

*(i) Defence pacts and neutrality accords*

Malta's graduation to statehood in 1964 occurred in a very uncertain environment, with serious doubts predominating about its economic viability, the reality or otherwise of its independence since it was still economically dependent on the United Kingdom and many misgivings about its ability to maintain its security without external help. Malta's foreign policy immediately reflected these concerns. Behind Malta's treatment of the 'hard security' problem lurked the financial question of how to use Malta's strategic advantage and excellent military facilities to earn adequate financial aid to strengthen Malta's economic development. Initially, these 'hard security" concerns were temporarily taken care of by a ten-year Defence Treaty with the United Kingdom (1964–74), which also included a financial aid package, allowing Britain to operate military bases in Malta, while giving the Maltese government some time and finance to grapple with economic development aiming to gradually wean off the Maltese economy from over dependence on British defence spending. A United Nations mission predicted, prior to Malta's independence, that this would take from 15 to 25 years.[26] The urgency of this task grew when the British decided to curtail defence spending and commitments in the 1960s. The British government's retrenchment policy in the defence sector was mainly directed at the UK's commitments East of Suez.[27] However, cuts were made across the board, including a shrinking of the UK's military presence in Malta, with obvious negative economic repercussions on Malta's economy, already suffering as a result of Britain's economic troubles of the late 1960s and successive sterling crises.

In 1972, the socialist government led by Dom Mintoff renegotiated a new Defence and financial pact with the UK and NATO, which included among other things a sixfold increase in the rent paid by the UK for the use of military facilities in Malta and a specific date, end of March 1979, when the military bases would close down. At an early stage in the dispute, the British government stopped payment of financial aid to Malta; however the Maltese government was able to obtain unspecified financial aid from Libya.[28] Concurrently, diplomatic relations were established with the People's Republic of China and an aid agreement was signed between Malta and China in Canton in 1972, providing for an interest-free loan to Malta amounting to 16.93 million pounds sterling. From that point onwards a number of large infrastructure projects, including the building of a new dock, were carried out by China. Relations with China remained strong even following the change in government in 1987, though a slow down occurred following the suppression of the Chinese democratic protests in Tiananmen square in June 1989. However, by the mid-1990s, relations began to warm up again and in October 1995, Chinese Prime Minister Li Peng visited Malta, underlining the strength of the relationship. The beginning of relations with the PRC had a number of spin-offs for both sides. The PRC managed to eject Taiwan from Malta, while the latter secured a substantial amount of development aid. Friendship with the PRC was used by Mintoff as yet another trump card in his dealings with the West. It secured him the aid and backing of a big socialist country, while geographic distance ensured that Malta could not fall easily under its hegemony. The move dispensed with the need to turn to the USSR as a counterbalance to the West, with whom relations were started only about a decade later. It has also been speculated that the attempt by Mintoff to introduce the Mediterranean issues in the Helsinki process may have been intended to disrupt the process on behalf of China, which feared that détente in Europe would strengthen the USSR against it by freeing her to transfer military forces from European theatre to its frontier with China.[29]

As 1979 drew closer, it became clearer that the closure of the bases and the forfeiture of rent earnings would have a negative effect on Malta's economy. Just prior to and following the closure of the military bases in March 1979, the Mintoff government sought political and economic sustenance of Malta's declared policy of neutrality, based on non-alignment. The government's aim was to continue to earn foreign rents to replace the ones which were being lost by the closure of the bases, this time by posing that it was rendering a valuable service to all its Mediterranean neighbours, and to regional security, by denying the facilities to all comers. Initially, a multilateral arrangement involving Algeria, Libya, France and Italy guaranteeing Malta's neutrality and providing a financial package looked in the offing. However, in the end only Italy signed such an agreement, recognising Malta's neutrality, while giving some military guarantees and providing financial aid to the tune of 12 million dollars a year for the period 1979–83. Apparently the other three countries' reaction was that once Malta had declared itself to be neutral and non-aligned there was no need for them to guarantee its status or to compensate Malta financially for adhering to this new position. An agreement combining recognition of neutrality with trade was also concluded with the USSR in 1981. This agreement raised a number of reactions, especially in Italy, as will be seen below. The trade agreement led to no substantial improvement in trade between Malta and the USSR.

After 1987, the Nationalist government showed no interest whatsoever in maintaining this policy of seeking both the recognition of Malta's neutrality as well as the financial aid for Malta to sustain such a policy, though it did not completely eschew neutrality. Addressing the diplomatic corps in June 1987, Prime Minister Adami declared that it was 'a geopolitical fact that in this day and age Malta's strategic position as an "unsinkable aircraft carrier" is far less important than in the past. In this respect, therefore, Malta's decision not to host any foreign military base (after 31 March 1979) was neither a gain nor a loss to any foreign country. . .'.[30]

The Prime Minister further elaborated that it was the intention of the government to abide strictly by the provisions of the Constitution in which neutrality was closely defined and was *sui generis*. This became the official working definition of neutrality and fitted well with Malta's policies as long as it was pursuing the EU membership application. Of course the definition is a minimalist one and the scope and definition of neutrality could be still stretched further by a government bent on so doing. The new government elected in 1996, promising to make Malta the 'Switzerland of the Mediterranean', has declared that it will pursue a policy of strict neutrality, adding that EU membership went against Malta's brand of neutrality. Malta has also insisted that Malta's neutrality is 'site specific',[31] i.e. related to its geographic position in the Mediterranean, the latter seen as being a confrontational area between North and South. No doubt Malta will be eyeing closely developments in Europe and especially the experience of Europe's neutrals in the European Union as possible role models.

*(ii) The approach to the European Community*

The other way in which Malta has attempted to strengthen its economic stability has been by expanding trade links. It was recognised at a very early stage (1950) that economic development had to overcome the smallness of the domestic economy by gaining access to external markets[32] and that it was more economical to target a group of states to achieve this.[33] The choice in the early 1960s fell between targeting the European Economic Community (EEC) and the European Free Trade Association (EFTA), of which the UK was a member. In the light of Britain's application to join the EEC in 1961, Malta declared that it wished to join the Community and subsequently sought (unsuccessfully) to conclude an Association agreement with the EC prior to independence. However, when the UK's application was rejected, Malta's interest in the EEC faded.[34] In 1967, Malta re-approached the Community with a formal application to conclude an Association Agreement. After an interval of about three years, such an agreement, providing for two stages, each for a period of five years and envisaging the eventual creation of a customs union, was finalised in 1970, becoming effective in

April 1971.[35] Although eventual membership of the EEC was not specified in the agreement, the Maltese government saw it as a step towards it.[36] The agreement was developed and expanded in the 1970s, through the addition of financial protocols and more concessionary rules of origin, in line with the Community's policies towards the non-Member countries of the Mediterranean region (MNCs). Pomfret has shown that the Association agreement has had positive economic effects on Malta's economic growth in the 1970s especially as regards its impact on the flow of foreign direct investment (FDI) and the rapid export growth.[37] Further, the agreement facilitated the diversification of Malta's export market, from over dependence on the UK market. By 1996, Malta's main export markets were in this order of importance: Italy, France, Germany, the UK, the Netherlands and Belgium.[38]

Relations with the Community entered a critical stage in 1980 when the first stage of the Association agreement came to an end and Malta demanded a "Special Relationship" that would comprise the main benefits of membership, including the right to use the structural funds, without the burdens. This relationship would have required the wholesale revision of the Rome Treaties to provide for a special category of 'second class' membership – besides raising serious precedents for the Community's relations with third countries. It was therefore resisted by Brussels, which nevertheless continued to renew the concessionary trade arrangements unilaterally despite the fact that the legal basis of the relationship had lapsed.[39] Relations with the Community began to normalise after the change in government in Malta in 1987 and culminated with the membership application in July 1990, the publication of the Commission's opinion on the application and the 'Structured Dialogue' with the EU, comprising a series of legal and economic reforms to prepare Malta for eventual membership.[40]

The approach to the European Community was central in Malta's foreign policy primarily because it was a means by which it sought to strengthen economic development. It was also the most economical way to gain quick and unproblematic access to foreign markets. One agreement with the EC was enough to open the doors to a large enough market in which to pursue diversification of products and outlets. On the other hand, a series of bilateral trade arrangements with non-EC countries based on Most Favoured Nation treatment (MFN), especially those with neighbouring Mediterranean states, made no substantial impact on Malta's trade patterns.

## MAINTAINING GOOD NEIGHBOURLY RELATIONS

The potential threat to the security of a micro-state comes from many quarters since every state larger than it has the power to use coercion, threats or aggressive actions in an attempt to influence its behaviour. Malta has tried to maintain good neighbourly relations with all the countries in its subregion but primarily with Libya, Tunisia and Italy. Libya has expressed concern that military facilities in Malta would be used against it. Italy has been consistently worried that Malta's independence and later the winding down of the British military bases would not lead to a vacuum which would be filled by the influence of some anti-western country.

### Libya

Malta established diplomatic relations with Libya in 1965 and opened an embassy in Tripoli in 1967. A trade agreement based on Most Favoured Nation (MFN) treatment as well as an economic, scientific and technical cooperation agreement were concluded in May 1968. Malta was one of the first countries to recognise the new Libyan government after the September 1969 coup which removed King Idris from power.[41] After the change in government in Malta in 1971 the approach towards Libya changed from pragmatism to one in which Malta stressed identical international aims. From 1971 to 1987, relations with Libya were used as a card in Malta's diplomatic games with the West. At the end of a visit to Libya by Mintoff in April 1972, the two sides affirmed their attachment to the principle of non-alignment and agreed to continue to ensure a Mediterranean for the Mediterranean peoples and free from the armed fleets of the superpowers.[42] On the first anniversary of that communiqué the two sides began contacts on the median line separating the two countries' share of the continental shelf.[43] This dispute kept simmering under the surface until August 1980 and the SAIPEM II affair, when the Maltese-contracted vessel by that name prospecting for oil in the channel separating Malta and Libya, was stopped from drilling by Libyan naval units. This threat, the first military one to Malta since independence,[44] prompted the Maltese government to dub Libya 'a danger to peace in the Mediterranean',[45] maintaining that the whole incident had been needlessly provoked, for the two sides had already agreed to take the dispute to the International Court of Justice in 1976.[46] On 20 March 1982[47] both sides exchanged ratifications to go to the International Court. A rapprochement was quick to follow the Court's judgment and a Friendship Treaty was concluded between the two countries in 1984 committing both sides to a 'continuous exchange of information on matters of special interest to the mutual security and defence purposes of the other side' and for Libya to train Maltese military personnel and to supply military equipment.[48] Both sides agreed to harmonise their views on political, economic, security and international issues affecting their interests and to support each other's viewpoint whenever such support is required.[49] More relevant was Article 3 of the Treaty, whereby Libya undertook to 'assist Malta whenever the Government of the Republic of Malta explicitly requests so in cases of threats or acts of aggression against Malta's territorial integrity and sovereignty.' It has been suggested that the clause could facilitate Libyan intervention in Malta in response to threats originating internally or following a request by a political force prepared to request its assistance.[50]

The hijack of an Egypt Air plane to Malta by the Libyan backed Abu Nidal group in 1985 is one testimony of the identity of views which the governments of the two countries shared. During the 15 April 1986 American air raid on Tripoli, Malta warned Libya of the impending attack some 45 minutes before it actually began, when allegedly some of the hostile aircraft were spotted on Malta's radar screens.[51] This was later confirmed by Gaddaffi himself in comments to Italian journalists, adding that Malta had in turn been warned by Italy of the attack.[52]

From 1987 onwards, relations with Libya were not downgraded, but Malta clearly showed that it wished to improve ties with the other Arab countries[53] – something which was hindered by the suspicions raised in some Arab capitals by the close Malta/Libya entente. New efforts were made to improve relations with Egypt, the Maghreb countries and Iraq. Military cooperation with Libya was discontinued when the government of Malta informed Libya in 1987 that it would not honour the relevant clauses of the 1984 Friendship Treaty because they were akin to a military alliance and that Malta wanted good neighbourly relations with Libya, but would not accept a position of subservience to her.[54] In 1990, a new treaty with Libya was signed in Valetta which dropped all reference to military cooperation.[55] The economic, trade and cultural aspects of Libyan–Maltese cooperation, stressed during the visit to Malta by the Libyan Prime Minister Montessori in October 1988,[56] have dominated bilateral relations since. For a long time Libya supplied Malta with petroleum at concessionary prices, though this was discontinued after 1988, while a number of manufacturing companies and hotels have been set up in Malta with Libyan finance.

This approach towards Libya was rendered difficult by domestic political considerations. The opposition Labour Party (MLP) at times attempted to exploit opportunities which arose. In December 1988, amidst rumours that the Americans were preparing an attack on an alleged chemical weapons plant in Libya, the MLP called for an urgent parliamentary debate on the situation and for Malta to raise the issue in the UN Security Council. The downing of two Libyan fighter planes in the central Mediterranean by US aircraft in January 1989 was followed by increased MLP pressure on the government, which was resisted. Malta adopted a neutral approach, limiting itself to urging Libya and the USA not to provoke each other and not to resort to armed conflict to resolve their differences.[57] During the 1989 crisis, the Leader of the Opposition travelled for consultations with the Libyan authorities in Tripoli following which he declared that the Libyans were not pleased with the position taken by the Maltese government.[58]

When the UN imposed sanctions against Libya in 1993,[59] Malta adhered to them. The Labour Party's reaction was that sanctions against Libya should not be strengthened.[60] Later, when in power, the Labour government declared that though it was not in favour of the sanctions, as a member of the UN, Malta would still abide by them. A ferry service, not covered by the sanctions, operates regularly between Malta and Libya. This has stimulated the growth of a lucrative trade with Libya, with Malta being used as a springboard, as well as a point of departure/arrival for Libyan nationals wishing to travel onwards by air from Malta, since flights in and out of Libya are banned. During 1995, three agreements were signed between Malta and Libya, one of which was a double taxation agreement originally concluded in April 1985.[61] The USA has objected that Libyan investment was taking advantage of the double taxation agreement between the USA and Malta, to bust the UN sanctions. The US–Malta double taxation agreement was unilaterally cancelled by Washington, pending negotiations on a new one that would presumably cover the loophole. The decision could possibly affect Malta more negatively than Libya, by discouraging American investment and penalising Maltese recipients of US pensions.[62]

The worse crisis in Libyan–Maltese relations occurred in 1995 following the assassination in Malta on 26 October 1995 of Fathi Shaqiqi, leader of the Damascus-based Islamic Jihad, who was travelling in transit through Malta on a Libyan passport. Libya suspended the ferry service to Malta for more than three weeks, while the Libyan authorities made it plain that the affair had soured relations between the two sides. A non-violent protest outside the Maltese embassy in Tripoli ended with a message in Arabic being handed over to the embassy threatening unspecified action against Malta if Shaqiqi's killers were not apprehended.[63] The Under Secretary in the Libyan Ministry of Foreign Affairs, Abdul Ati Al-Obeidi, visited Malta to discuss the incident while the Maltese Police Commissioner visited Tripoli to discuss ways in which such incidents would not happen again.[64] However, relations slowly returned to normality by the end of 1995 when it became clearer that the Maltese authorities, unaware of the identity of Shaqiqi and the security risk he presented, could not be held responsible for what had happened.[65] Neither did the episode negatively affect newly revamped diplomatic relations between Malta and Israel,[66] despite the fact that Libyan pressure was reported to have been exerted on Malta to end these ties.[67]

Libya's attitude towards Malta's membership of the EU is not negative, and it has shown a degree of readiness to concur with it, as long as there is no conflict of interests, as was stated by the Libyan Foreign Minister, Omar Mustafa al-Montessori during a visit to Malta in August 1995.[68] The former Maltese foreign Minister had stated that the two sides have no conflict over the EU membership issue adding: 'Obviously, Libya has its national interests to safeguard and Malta has its own national interests'.[69] However, Libya has shown reservations regarding Malta's participation in NATO's Partnership for Peace Programme (PfP). This issue too had been discussed at length with Mr Montessori and again the former Maltese foreign Minister had stated that 'the Libyan Foreign Minister wanted to have an assurance that PfP and NATO were not one and the same thing. I explained to him they were not and the impression I got was that he fully understood our explanation'.[70]

Following the 1996 election, Malta left the Partnership for Peace programme – ending a relationship which was of concern, though hardly a real threat, to Libya. The new Maltese Foreign Minister described the move as not being 'a political measure aimed at distancing Malta from Europe'.[71]

**Italy**

Malta's history has been closely linked with that of Italy for centuries. The possibility of strong Italo-Maltese relations featured prominently in the recommendations on trade and economic development given to the Maltese government prior to independence. In a wider arrangement envisioning membership of EFTA and a customs union with the UK, Malta was urged to seek 'special arrangements' with Italy, to become 'part of the Italian economic sphere" in order to attract Italian investment in manufacturing and tourism.[72]

In the period 1962–3 when Malta entertained the possibility of signing an Association Agreement with the EC prior to achieving independence, consultations took place in Rome on the possibility of Italian support to help Malta achieve this aim.[73] Italo-Maltese relations and collaboration on Malta's quest for EU membership strengthened after the establishment of diplomatic relations in 1965. Italy offered the services of an expert to help Malta prepare its approach to the Community. Worried about the defence implications of Britain's decision to curtail its military commitments in Malta as a result of defence spending retrenchment of the mid-1960s, reference to which has already been made, the Italians had also offered financial aid to compensate the losses Malta stood to make as a result of the British defence policy.[74] A number of agreements were signed with Italy in the period 1967–8 covering cultural, technical and scientific co-operation, financial aid and trade.[75] While Mintoff was negotiating the new Defence Agreement with the UK (1971–2), Italy played a very positive role towards a successful outcome and subsequently signing a financial and economic co-operation agreement with Malta, providing 1,500 million Lire.[76] An Italian technical military mission has also been in Malta since 1973, helping in various infrastructure projects. As 1979 was approaching (the end of British military facilities in Malta) Italy unsuccessfully attempted to use its EC membership to try and convince France to join it in recognising Malta's neutrality and provide financial support. A multilateral guarantee of Malta's neutrality cum financial aid involving Italy, France, Libya and Algeria failed to come through. However, following the 1980 SAIPEM II affair with Libya, the neutrality agreement was signed with Italy.[77] This agreement also included a financial protocol. Since then, four financial protocols have been signed; the last covering the period 1995–2000.[78]

When the Italy–Malta neutrality agreement was presented for ratification to the Italian Lower House on 10–11 March 1981,[79] it came under sever criticism primarily because Italy had done it alone bilaterally, when originally there had to be a multilateral treaty, which would have made more sense given that the stated aim of the Italo-Maltese Pact was to contribute towards the enhancement of peace in the central Mediterranean region. Further criticism made was that Italy had not consulted its NATO allies before committing itself to military intervention in case of a violation of Malta's sovereignty and that this was not automatic but depended on the exclusive judgement of the Italian government. The conclusion by Malta of another neutrality agreement with the USSR in 1981 and the Friendship Treaty with Libya in 1984, coupled with the deteriorating political and economic situation in Malta after the 1981 election heightened disenchantment with Malta's policies in Rome and part of the funds promised under article III of the Financial Protocol of 1980 were withheld. The second financial protocol was signed in 1986, but no funds were transferred to Malta pending ratification of the accord by the Italian Parliament. Strong diplomatic pressure by Malta failed to make headway and dispatches to the Ministry of Foreign Affairs by Rome-based Maltese diplomats show that every time they raised the question of the financial protocols, discussion was quickly diverted to the political situation in Malta. When the protocols were eventually debated and approved by the Italian Parliament in 1988, again frequent reference was made by Italian deputies to the Maltese socialist government's domestic and foreign policy prior to the 1987 election.

Italo-Maltese relations improved considerably after 1987 and significantly Malta's application to join the European Union took place during Italy's turn at the Presidency of the EU. The Italians supported Malta's application strongly, but then due to the effects of the internal political situation in Italy and the *tangentopoli* affair,[80] factors which reduced Italy's effectiveness at the European level, Malta's application was somewhat neglected. The situation began to improve in 1995 following the Italian elections and the formation of the government of Prodi. This was most evident at the Florence European Council meeting of June 1996 when the Italians pushed unsuccessfully for a more tangible boost to Malta's EU application. However, the new Maltese government's decision not to attend the Dublin Council meeting six months later elicited public shows of disappointment from Italy's prime minister. It is more than likely that geographic proximity and the political, cultural and economic ties which have linked the two countries for centuries and which have been resilient both to the frequent changes of government in both states will ensure a continued strong relationship between the two sides, even in the event of Malta not joining the EU.

## RELATIONS WITH OTHER COUNTRIES OF THE MEDITERRANEAN REGION

Besides relations with its two closest neighbours, Malta has established relations with most of the other countries of the Mediterranean region. Relations with the Arab Mediterranean countries led to the signing of bilateral trade agreements mostly based on the Most Favoured Nation clause. Trade did not however expand briskly with these countries as more than three- fourths of Malta's commerce remained directed towards the European Community and only around 2.5 per cent goes to the Mediterranean non-EU countries, despite efforts to diversify markets. In the period when Malta-Libya relations were very solid, relations with other Arab states in the region were 'crowded out' by the intensity of the former. From 1987, efforts were made to balance relations with the Arab world and an embassy was opened in Cairo and later in Tunis. On the diplomatic front, Malta had close contacts with a number of Mediterranean countries in the context of the Non-aligned Movement, the Barcelona Convention, the Western Mediterranean Forum, the International Parliamentary Union (IPU) and the Mediterranean Forum. Similar avenues of dialogue with the Mediterranean countries exist now in the framework of the Euro-Mediterranean Partnership initiative of the European Union. The follow-up conference to the Barcelona meeting of 1995, took place in Malta in April 1996 but ended inconclusively.

At the bilateral level, Malta has a 20–year-old dispute over the delineation of the continental shelf with Tunisia akin to the one it had with Libya. The problem came to a head in 1994 and preliminary discussions began between experts from the two countries in the Tunisia–Malta Mixed Commission.[81] It is worth recalling that Malta's territorial sea is already 100 times its land area and could be 200 times bigger once the outstanding disputes are resolved.[82]

## PARTICIPATION IN INTERNATIONAL ORGANISATIONS

A small state gives relatively more importance than larger ones to international organisations. Besides world organisations such as the United Nations and the non-aligned movement, there are two other types of organisations or multilateral initiatives in which Malta has participated since independence: those in Europe and those in the Mediterranean. In the case of Europe the most important ones are the Council of Europe, the European Union and the Conference on Security and Co-operation in Europe (CSCE) later the OSCE. In the case of the Mediterranean these include action under the Barcelona Convention, the Western Mediterranean Forum, the Mediterranean Forum and the Euro-Mediterranean Partnership. Malta also came up with two proposals for the Mediterranean: a Council of the Mediterranean and a Mediterranean Stability Pact launched in March 1995 and which is being pursued within the framework of the Euro-Mediterranean Partnership launched in Barcelona in November 1995. Activity in these multilateral organisations was initially considered as a means of affirming Malta's independent identity and subsequently as the best available tool for upholding international norms and law. They also presented useful avenues for obtaining support on Maltese international initiatives and as a means for trying to influence in concert with others, the international environment and reduce risks for Malta.

### The United Nations

Membership of the United Nations[83] was seen as an affirmation and recognition of Malta's statehood. Addressing the UN General Assembly in 1967, the Maltese Prime Minister declared: 'The weaknesses we share in size (with other small states), wealth and power give us a natural interest in the United Nations'. He added that Malta would pursue a four-pronged programme in relation to the UN, namely disarmament, peaceful settlement of disputes, peace-keeping and the economic and social work of the organisation.[84] It was at the United Nations that Malta embarked upon a number of initiatives most central of which was the historic statement made by Ambassador Pardo on 1 November 1967 to the first committee of the UN,[85] proposing a new law of the sea based on the 'qualitative' concept that the oceans and the seabed beyond national jurisdiction be declared the 'Common Heritage of Mankind'.[86] The initiative was intended to head off a struggle of the strongest in the exploitation of ocean resources beyond national jurisdiction by creating a new and equitable law of the sea. It was also intended to facilitate the peaceful resolution of disputes over the delineation of territorial seas and economic zones especially in very closed seas such as in the Mediterranean where for example, a 200–mile economic zone made little sense. It proposed giving landlocked states a share in the riches of the oceans and ultimately it was a concept to favour small and weak states against the strong. Malta also made a number of proposals (1968) on disarmament, the most crucial of which was that action needed to be taken at world level against the production of chemical and bacteriological weapons, which did not depend on big science and big technology, and hence was accessible to any state which could acquire a fertiliser factory.[87] A year later Maltese Minister Vincent Tabone formally raised the question of the aged in the General Assembly and two years later Malta introduced and piloted the resolution on the 'Question of the Elderly and the Aged' which was adopted by the 26th session of the Assembly. These initiatives, and most pointedly the one concerning the new law of the sea, were abandoned by the Mintoff government soon after it was elected in 1971.[88] Rather belatedly, after loosing the initiative on the law of the sea for more than a decade, the Mintoff government attempted to attract the headquarters of the Sea Bed Authority to Malta but failed. The most negative effect of this neglect was not however the loss of the Authority but the fact that Malta had lost the pre-eminent position it enjoyed in the negotiations and was powerless to influence developments especially on sections of the Convention which would be detrimental to Malta, as for example the principle to be employed in the delimitation of the median line in continental shelf disputes.[89] In the dispute over the continental shelf with Libya, Malta was as it were, hoisted by its own petard.

Unlike the Nationalist Administration of the 1960s, the Maltese socialist government gave lesser importance to the United Nations. In the period 1971–87, no Maltese socialist Prime Minister addressed the General Assembly of the United Nations. When the Nationalist Prime Minister Adami addressed the Assembly in 1987 he reiterated that the same principles which guided the Nationalist government of the sixties in its participation in the UN would likewise guide Malta from there onwards. Malta's return to a new activism in the United Nations was the reinstatement of the basic line first adopted in the sixties that 'mini-stature does tend to make our self-interest more nearly coincide with the global interest' and that a diminutive physique goes with a point of view that is singularly alert to the prospects and problems of the survival of all in a danger-filled environment.[90]

**The Non-Aligned Movement (NAM) and the neutrality non-alignment nexus**

In the period 1964–71, the Non-Aligned Movement was shunned by Malta. However, it gradually became of central importance in the government's rhetoric and external relations after 1971, especially as it sought to exploit ColdWar rivalries. From 1987 onwards, there followed a marked departure from this policy. Addressing the second meeting of Foreign Ministers of the Mediterranean Members of the Non-Aligned movement in June 1987, Malta's Foreign Minister, Censu Tabone, declared that Malta would not subscribe to the Movement's anti-Western rhetoric.[91] By 1989 as the crisis set in the movement over its *raison d'être* following the end of the Cold War, Malta was urging a reform of the movement and a reassessment of its role.[92] Malta's membership of the movement became questionable in the context of its EU membership application, considering that Malta's status of neutrality and its policy of non-alignment have been singled out as factors which may cause difficulties to the Union's Common Foreign and Security Policy (CFSP) in the event of Malta joining the Union.[93] The present Labour government does not appear enthusiastic about non-alignment, though it participates fully in its meetings and while declaring that EU membership would have negative effects on Malta, says that in the framework of its alternative, that of establishing closer and stronger relations with the European Union but falling short of membership, it can negotiate a bilateral agreement on security with the Union. The MLP's proposal envisages that the agreement would provide that on the sole request of the Maltese government, 'when this perceives a foreign threat of violation, or the violation of the sovereignty, neutrality, unity or the territorial integrity of the Republic of Malta, the European Union undertakes to help Malta face the situation with the best means... including military assistance'.[94] The Labour Party has stated clearly that EU membership would undermine Malta's neutrality.[95]

The link between neutrality and non-alignment[96] is established in the Maltese Constitution which defines Malta as a neutral state, adhering to a policy of non-alignment and refusing to participate in any military alliance.[97] Considering the declining importance of the Non-Aligned Movement and especially the weakening and disappearance of the European group within it following the break-up of Yugoslavia and eventually Cyprus's accession of the EU (not to mention Malta's own possible membership if a change in present policy occurs), it is unlikely that participation in the Movement will reach the high level of importance in Malta's foreign policy that it had in the 1970s and 1980s.

## OTHER EUROPEAN INSTITUTIONS

Malta has participated in two other European organisations of importance namely the Council of Europe (CoE) and the Conference on Security and Co-operation in Europe (CSCE), later the OSCE.[98] Membership of the CoE was one of the ways by which Malta affirmed its European character following independence, since it was the only pan-European organisation which it could join with relative ease and at the same time acquire a means of participating in European affairs. Malta's record in the CoE is not however unblemished. Furore was created by Mintoff when addressing the otherwise docile Parliamentary Assembly in 1978, in his capacity of Chairman of the Committee of Ministers, when he made a violent attack on Western European policies and values, accusing it of lacking an identity. He took umbrage at the fact that Western Europe had failed to come up with military and economic guarantees for Malta's neutrality.[99] Relations with the Council of Europe deteriorated from there onwards and after the anomalous 1981 election result in Malta which saw the Labour Party gaining more seats in the Maltese House of Representatives with lesser votes than the Nationalist Party.[100] A number of issues were raised in the CoE in the period 1981–87, primarily the human rights situation, the attempt to take-over Church schools, the confiscation of church property and the 'Foreign Interference Act', by which the socialists attempted to gag the opposition and cut them off from international support.[101] Between 1983 and 1985, Malta failed to appoint a parliamentary delegation to the CoE Parliamentary Assembly and was thus boycotting the sittings.

Malta's participation in the CoE returned back to normal after 1987. The first concrete step was taken when substantial parts of the Human Rights Convention, hastily ratified by the socialist administration just prior to the 1987 election, was incorporated into Maltese law.[102] In May 1988, Malta's Prime Minister was invited to make the opening address of the Congress of Europe held in The Hague to mark the 40th anniversary of the Congress which paved the way for the setting up of the Council of Europe. In 1990 Malta was proposing that the Council of Europe take an active part in the harmonisation of social policy in Europe while concurrently bringing Mediterranean issues to the attention of the Parliamentary Assembly.[103]

## The CSCE and the Mediterranean policy

One very interesting European forum in which Malta has participated is the CSCE, now OSCE, in which it has promoted the so called Mediterranean chapter, linking European Security with security in the Mediterranean region. The idea that security in the Mediterranean should be discussed in the Helsinki process was originally put forward by Algeria and Tunisia in 1972 when they circulated a note in the conference proposing that they should be allowed to participate in some way. Their cause was temporarily taken on by Italy. The two sides of the East–West divide were however reluctant to discuss the issues of the Mediterranean as this would surely introduce the very divisive Palestinian Question and increase the likelihood that the conference would fold up in complete failure without achieving any progress on the wider issues threatening European security.[104] Despite the peripheral nature of Mediterranean stability to the aims of the Conference, Malta paid little heed to the wider concerns and concentrated on this issue closer home insisting that the final declaration of Helsinki include language on reducing the US and Soviet fleets in the Mediterranean. In the end, after the Mintoff government blocked the conference – since decisions in the CSCE are taken by consensus – agreement was reached on making a reference to the reduction of forces in the Mediterranean region.[105] This initiative led to a follow-up meeting of experts in Valletta on cooperation in the Mediterranean region, which however failed to achieve any substantial progress.[106] The meeting simply established that cooperation with the non-participating Mediterranean countries could be better served under the aegis of UN institutions such as UNESCO, UNEP and the Economic Commission for Europe (ECE).[107] Four years later, Malta tried once more when it attempted to extract further concessions on Mediterranean issues at the Madrid follow-up conference by holding up the approval of the concluding document. However, when it became clear that all the other participants would proceed, with or without Malta, the latter finally relented.[108]

Following Madrid, Malta worked from within the group of neutral and Non-Aligned countries in the CSCE process with no major initiatives that had characterised its participation in the first decade of the CSCE. Indeed, the Helsinki process was the only multilateral process where Malta grabbed the opportunity to push forward its Mediterranean policy. It was to be expected that in its foreign relations, a micro-state would emphasise regional concerns, in this case that peace and stability in the Mediterranean would be given priority and undivided attention. However, Malta's policy of insisting on the departure of the superpower fleets from the Mediterranean could be explained in two basic ways: either that Malta failed to take account of the wider international context or that it pursued a deliberate policy intended more to undermine American interests in the region than those of the Soviet Union. Many countries of the region including Israel, Egypt, Morocco and Tunisia – not to mention the NATO member states of the northern littoral – would have been loathe to see the departure of the American Sixth Fleet considered by them a factor of regional stability. Malta's policy was on the other hand consistent with that of Libya on the matter. The result was that Malta was isolated in its Mediterranean policy from the rest of the Mediterranean countries on whose behalf its foreign policy was allegedly being proposed.

## A FUNCTIONALIST APPROACH

Malta's Mediterranean policy in the period 1971–87 was an ideologically loaded one which pleased only one country in the region. The lack of widespread support made it irrealisable. The problem was compounded by the fact that it was opposed by the superpowers as well. It was anti-Western for, if realised, it would have entailed the departure of the Americans, but not of the Soviets, who by virtue of geography could claim that they too were a Mediterranean country, besides continuing to enjoy right of passage from the Black Sea to the Atlantic via the Mediterranean. Its anti-Western bias was more apparent by virtue of the fact that, while Malta insisted on the departure of the fleets of the superpowers, it signed a neutrality accord with the Soviet Union in which both sides expressed their readiness to consult one another on 'questions directly affecting the interests of the two countries, including the neutral status of Malta, and in case of situations arising, which create a threat to peace and security or the violation of international peace, will also be prepared as necessary, and at the request of the Government of the Republic of Malta and by agreement between the Parties, to enter into contact with [one another] so as to co-ordinate their positions in order to remove the threat or to establish peace.'[109]

The alternative to this policy was presented by the Opposition Nationalist Party (NP) in 1983.[110] Basically, the NP held that the naval presence of the superpowers in the Mediterranean region was purely symbolic, to communicate messages, a display of force more than anything else. The Mediterranean was described as a fragmented region, full of conflicts and at the same time a transitional zone between the developed North and the developing South. It was useless for a small state to make quite a lot of noise on issues when it was certain it will fall on deaf ears. An alternative course of action was suggested such as promoting the development of a functionalist approach as originally formulated by Mitrany[111] to the solution of the region's problems. This would also help the countries of the region skirt around the difficulty created by the fact that the northern littoral states were integrating together while the southern shore countries aspired to the unification of the Arab nation. As a model of this cooperation the NP leader cited the Barcelona Convention on the protection of the Mediterranean sea against pollution signed in 1976.[112] That year, under the aegis of the Convention, a regional centre to combat oil pollution and other harmful substances in the Mediterranean was established in Malta. Four areas of cooperation were proposed by the NP: joint exploitation of mineral resources, the development of unconventional energy sources, fishing and the protection of the marine environment. Cooperation in these sectors would then spill over into other areas. Of course as neo-functionalists would argue, such a process also required political impetus to move forward.

From 1987 onwards the Nationalist Party in government had the opportunity to develop a policy towards the Mediterranean on these 'functionalist' lines. What actually resulted was the opposite. Malta participated in or enthusiastically supported initiatives such as the 'Five + Five' Dialogue in the Western Mediterranean, the proposal for a Conference on Security and Co- operation (CSCM) in the Mediterranean and the Mediterranean Forum. The 'Five + Five' encounter was initiated by France and was intended to contribute towards the stabilisation of the Maghreb region, an area of most concern to France. It involved Algeria, Morocco, Tunisia, Libya and Mauritania on the Arab side and France, Spain, Italy, Portugal and later Malta on the European side. Hopes were kindled that since the forum did not include any of the direct belligerents in the Middle East problem that it would actually succeed and eventually serve as a model for the consolidation of stability in the rest of the region. These hopes were dashed by the worsening situation in Algeria and by developments on the Lockerbie affair and UN imposed sanctions against Libya. The CSCM was an Italo-Spanish initiative, supported by the EC, at the CSCE conference in Palma de Mallorca on 24 September 1990. The Mediterranean Forum was a joint French-Egyptian initiative started in Alexandria in July 1994 and from which Malta was at first excluded. The forum's beginnings owed much to the failure of the 'Five + Five' and more to the ameliorating situation in the Middle East.

As a Mediterranean country with obvious interests in maintenance of stability in the Mediterranean region, Malta could not but participate actively in the fora just mentioned. However, when it gave birth to its own initiative, in November 1992, Malta departed from its 'functionalist' approach to the Mediterranean to propose the setting up of a 'Council of the Mediterranean' based on the structure of the Council of Europe.[113] The proposal was made by the Foreign Minister Guido de Marco during a conference in Tunis organised by the *Rassemblement Constitutionnel Démocratique*.[114] The idea of the Council of the Mediterranean never got off the ground and when the Mediterranean Forum came into being, Malta was at first excluded though it was allowed to join later. The Nationalist Government never seems to have taken the Council of the Mediterranean under its wings and an article by the Maltese Prime Minister in the Winter of 1995, made reference to all Mediterranean initiatives, including the Mediterranean Forum, but not to the Council of the Mediterranean.[115] The PM, while referring to the processes endangering stability in the Mediterranean, and while giving weight to the need for dialogue, praised the idea of the CSCM 'because it promises to support a sufficiently broad and flexible agenda and to permit from the outset a properly structured participation by both Mediterranean and extra regional states in the process of building a Mediterranean dialogue.'[116] He concluded by saying that the dialogue which must be set up in the Mediterranean region 'must be able to adopt pragmatic approaches, especially in the search for and utilisation of initiatives for functional co-operation'.[117] The Malta Labour Party (MLP) is in favour of the CSCM, although it is inclined to giving greater importance to the so-called Mediterranean Chapter of the Helsinki accords which has to date remained a dead letter. It is also in favour of the creation of a Mediterranean regional forum, Council or Assembly.[118]

In March 1995, while addressing the Final Conference of the Pact on Stability in Europe, Malta's Foreign Minister advanced the idea of promoting a Stability Pact for the Mediterranean region. The Pact is to be based on eight principles:

(i) Refraining from the threat or use of force;
(ii) peaceful settlement of international disputes;
(iii) inviolability of frontiers and territorial integrity of states;
(iv) right of peoples to self-determination and to live in peace on their own territory within internationally-recognised and guaranteed frontiers;
(v) sovereign equality of states and non-intervention in internal affairs;
(vi) respect for human rights;
(vii) cooperation between states and
(viii) fulfilment in good faith of obligations assumed under international law.[119]

The idea was further elaborated by Malta during the Euro-Mediterranean Partnership conference held in Barcelona on 27–28 November 1995, indicating that the Mediterranean Stability Pact should be pursued through the Barcelona process. The likelihood of its success were therefore stronger than if Malta attempted to pursue the idea on its own. Further, the Stability Pact initiative seemed to enjoy the support of the new Maltese government, and in all probability, the initiative will be pursued further in the future.

## FUTURE DEVELOPMENTS

Malta's foreign policy in the rest of the nineties and the foreseeable future will continue to be based on the special characteristics of smallness and vulnerability both in economic as well as in security terms. It will therefore continue to seek to achieve its aims through multilateral fora and international organisations, not least among these the European Union, the Council of Europe and the United Nations. Of relevance too will be its regional concerns characterised by its interest in the stability and environment of the Mediterranean region and relations with its neighbours especially with Italy and Libya.

The most important aspect of its foreign policy will no doubt be the European Union. Negotiations for entry into the EU were due to begin six months after the end of the Intergovernmental Conference (IGC). The present Socialist administration has declared that membership of the EU is not to Malta's advantage in the foreseeable future, but at the same time it has not completely closed the door on membership. It is struggling to maintain the 'structured dialogue' begun by the EU with each of the applicant states even though this may have been rendered superfluous by Malta's decision to slow peddle on membership. Viewed as a security community, the EU provides Malta with guarantees to its further economic well-being (increased market access and scope for export diversification, trade creation, structural funds, monetary stability and the single currency) as well as with security from external challenges which abound in the region. The former Prime Minister of Malta speaking at the Paul Henry Spaak Foundation in Brussels in 1994 put it this way: 'We live in a dangerous sea. We have always lived in it. Yet today we have the opportunity of anchoring our country to an emerging political union which shares our moral values, our Christian culture and our beliefs in democracy, the rule of law and social justice. We see the EU gaining strength and cohesion, being endowed with its common foreign policy and tomorrow, perhaps, with its common security policy. Malta's security is the first aim of our foreign policy, and it is in our interest to belong to an entity which makes us more secure' [120] Economic prosperity on the other hand enhances internal stability and the security of Malta in general.

The Labour Party declares that Malta must have the closest and best relations with the EU in accordance with its interests as well as with those of Europe and Malta's neighbours,[121] the latter being a not so concealed reference to non-EC Mediterranean countries especially Libya. The MLP also declares concurrently that 'the party believes that our relations with

Europe should be primarily based on the considerations of Malta's national interest.' [122] Speaking at a Press Conference on Malta's EU membership just before the October 1996 election, the leader of the MLP, Alfred Sant, referred to a meeting he had had with Libyan leader, Muammar Gaddaffi in Tripoli in 1995 in which the Libyan leader allegedly informed him that he would be assessing whether Malta's relations with the EU would disadvantage Libyan interests as well as those of the rest of North Africa.[123] This position does not contrast significantly with the assurances given by Libya's Foreign Minister to his Maltese counterpart in 1995 that Libya was ready to concur with Malta's EU membership as long as there is no conflict of interests. The position adopted by the MLP that its foreign policy would be based on a policy which would coincide with the interests of its neighbours has been criticised as being a dependent variable of Libyan policy.[124]

In March the Foreign Minister Vella, referring to the frequent visits of warships from the United states and its NATO allies was reported to have told *The Times of Malta*, that the number of such visits was bound to decrease as a result of a note sent to all diplomatic missions in Malta that such visits should be for good reason and not simply 'for the sake of it'.[125] Libyan foreign Minister Al-Montasser, visiting Malta in April 1997, made an indirect reference to Malta's participation in the PfP and expressly to the question of visiting war ships when he told a news conference at the end of his visit to Malta that 'Our only worries [regarding Malta's relations with the EU] will be if policies are implemented with some military alliance which would give Libya some apprehensions' and further that the frequent visits of naval units to Malta make Libya 'a bit jittery'.[126] It was during the same conference that his Maltese counterpart said that if Libya was not included in the Euro-Mediterranean Partnership, 'the process would have to stop'.[127] Both sides called for the end of UN sanctions against Libya and stressed that Libyan involvement in the Maltese economy would be stepped up. Speaking at the Non-Aligned meeting in New Delhi a few days later, the Maltese Foreign Minister again called for an end to UN sanctions against Libya, a policy line which the Maltese government proceeded to reiterate a number of times in the days previous to the Euro-Mediterranean Partnership meeting in Malta.[128] On 24 April, the DOI issued a statement that, during a meeting of the Libya–Malta mixed Commission, agreement was reached on a number of outstanding bilateral issues, foremost among which a commitment by Libya to pay Malta all outstanding debts 'within a short time' and an agreement binding Libya to buy consumer goods from Malta worth between 15 and 20 million US dollars.[129] Clearly, a tilt towards satisfying Libyan demands in the area of Malta's foreign and security policy has occurred and the *quid pro quo* is rather clear and seems to have been established between Libya and the Labour Party while the latter was still in opposition.

The MLP believes that EU membership undermines Malta's neutrality based on non-alignment.[130] It holds out the model of a Switzerland of the Mediterranean, neutral and prosperous. It believes that neutrality helps Malta to achieve deeper relations with a larger number of countries than would otherwise be the case if it joined the EU.[131] For this purpose, as an alternative to membership it proposes that Malta concludes an industrial free trade agreement with the Union over an unspecified longer term period. At the same time, the MLP promises to remove VAT, introduced in Malta from the start of 1995, and to reintroduce the customs duties in place before VAT was introduced.[132] The introduction of VAT has facilitated the removal of customs duties, which had a large fiscal element in them, and the adaptation of the European Community's external tariff, a move consonant with Malta's membership aspirations but also with the alternative of achieving a free trade area with the EC. Though the removal of VAT became a major priority when the new government took office in October 1996, the implementation of an 'alternative' tax system was delayed due to the opposition of all of Malta's social partners, including the General Workers' Union (GWU) which closely supports the Labour Party. Even the International Monetary Fund (IMF) advised the government not to remove the VAT system. The latter has on the other hand taken a rather unpragmatic position of sticking with its electoral promises.

On the question of national security, the MLP believes that this is best safeguarded by Malta continuing to adopt a policy stance of neutrality based on non-alignment[133] even though the party admits that 'in the present geopolitical environment, should Malta ever be faced with any security threat, the probabilities are that these would be more of a *non-military* rather than a military nature.'[134] Four pillars for a national security policy are suggested: (i) together with other Mediterranean states Malta will seek to promote more resolutions on security in the Mediterranean in the United Nations, while working for greater recognition of the special needs of smaller members of the organisation; (ii) in the OSCE, Malta must actively promote the rights and protection of the sovereignty of small states, to strengthen the OSCE's institutions, especially peace-keeping, and to make the OSCE's activities more accessible to all Mediterranean countries; (iii) a bilateral security arrangement with the European Union and (iv) the strengthening of the non-military links with the Mediterranean non-EU states in the fields of the environment, commerce, finance, social issues and crime-prevention.[135]

The MLP envisages that the bilateral security arrangement with the EU should be based on the understanding that solely upon the request of the Maltese government, when this perceives a foreign threat of violation, or the violation of the sovereignty, neutrality, unity or the territorial integrity of the Republic of Malta, the European Union undertakes to help Malta to face the situation with best means then available at its disposal, including military assistance. This also applies when [Malta] has to exercise its right of legitimate defence.[136] Two problems can be associated with this proposal. The first is that it is not so clear that the EU will be willing, given the legal and logistic capacity, to give such military guarantees, if only for the precedent it will raise in its relations with third countries. The second is that if the defence element of the CFSP remains undeveloped, and if it is pushed further in NATO's hands, as appears to be the current trend, the EU will neither be in a position to deliver the goods nor to sign such an accord. Further, such an arrangement as is being proposed by the MLP would, if realised, further place Malta in a dependency relationship with the EU, while membership of the EU would not. Dependence would be unlimited and in its foreign policy Malta could eventually be reduced to a satellite status. It may attempt to resist this from happening and take a markedly different course, a policy of counter-dependence which would poison its relations with the Community. Counterdependence can be played only in the company of the EU's adversaries which in the Euro-Mediterranean Partnership context would mean pairing up with Libya, if the latter is so inclined.

From the standpoint of the economic consequences of Malta's two alternative approaches to relations with the European Union, the costs and benefits of each are not clearly different, though the present government says they are and that the benefits tilt in favour of a relationship which just falls short of membership. Both models entail further liberalisation of the Maltese economy but at different speeds and to different extents. The FTA model will either resemble the Euro-Mediterranean Partnership one, in which case the agreements concluded by the European Community with such non-member Mediterranean states as Israel, Tunisia and Morocco are instructive. Alternatively the FTA may be based on European Economic Area (EEA) agreement concluded between EFTA and the EC. The latter would entail that Malta would join the internal market and hence achieve a higher degree of integration with the EU than in the case where it would base the relationship on the Euro-Mediterranean partnership model.

Though presently Malta seeks a relation with the EU which falls just short of membership, it is possible that the government could eventually change its position since it has not retracted the membership application. The Labour Government appears to approach the question from the standpoint that a free trade area agreement is not only in line with its electoral promises but is quicker to achieve. The question as to which of the two alternatives before Malta is the most economically suitable is still unresolved as both sides of the political divide advance conflicting assessments. One of the major considerations that appears to tilt the argument in favour of membership concerns the funding which Malta stands to gain in terms of structural aid.

The other main question to be considered is political and has to do with Malta's overall security, in other words whether neutrality as such will guarantee Malta's security and whether the Swiss model is applicable to the particular situation existing in the Mediterranean. The Nationalist Government believed it does not, and so does the Labour Government which in the past (1980–81) obtained military guarantees for Malta's neutrality from Italy and the Soviet Union and now seeks to obtain similar guarantees from the EU. This mistrust of neutrality in Malta as an acceptable security guarantee results from the fact that, even barring Malta's experience as a micro-state which has frequently been occupied by a foreign power, history is permeated with instances when the neutrality status of small weak states was disregarded and violated by more powerful states. Nicholas de Flue's advice to the Diet of Stans in 1481 'do not get involved in the quarrels of others'[137] does not offer much assurance to the Maltese that others will not poke their noses in Maltese affairs, even though this appears to be the present government's stance. A rational model appraisal is enough to show that external challenges cannot be adequately deterred by neutral Malta. This leaves us with the second problem of how adaptable to the unique situation of Malta is the Swiss model of neutrality. Freymond rightly described Swiss neutrality as *sui generis*.[138] Bohn concisely outlined the underlying conditions which made neutrality a successful foreign policy for Switzerland.[139] Among many of the conditions he cites, the main is surely the advantageous military position of Switzerland from the point of view of both land form and internal means of communications. Few states aspiring to hegemony could have afforded the expenditure to dislodge them.[140] By contrast it is not so easy for Malta to defend itself and not so expensive for an envious power to attack it, as history has shown time and again. Neutrality served Switzerland well economically for a long time. It may no longer be the case and the Swiss are weighing the economic benefits of the trade-off between maintaining neutrality and joining the EU. For Malta, neutrality means forgoing the benefits of membership of the EU, which are both economic and political, though the current Maltese government argues that these advantages are bigger if Malta stays out

Frequently, the situation in the Mediterranean has been described by the Labour Party as one of competition between north and south: 'Malta must adopt a policy which prevents its transformation into a fortress on the European frontier guarding against the Arab world and vice-versa.'[141] However, the countries of the southern shores of the Mediterranean do not view the northern shore state as a military threat (except in Tripoli) any more. but as an essential partner whose intrusive presence is not feared but rather its lack of interest.[142] In a confrontational context neutrality would make moral sense, though it could make national security worse. However, when north and south are speaking and acting not in an adversarial manner but towards consolidating a Euro-Mediterranean partnership, neutrality makes little sense.

Malta's neutral status could still stand in the way of EU membership if this is the direction which is eventually taken. In the Commission's Opinion on Malta's membership application it has been stated that 'the Maltese Government's statement that it is in the country's interest to subscribe to the Common Foreign and Security Policy (CFSP) does not alter the fact that it might be necessary to amend the constitution if Malta is to participate fully in that policy as it develops over the next few years.'[143] Changes to the Constitution require a two-thirds majority. The Nationalist Government held that since four neutral countries, namely Ireland, Austria, Sweden and Finland are already members of the Union, no Constitutional amendments are required for Malta to participate in the EU.[144] The solution therefore lies in the way in which the Constitution will be interpreted. And from 1987, Malta's neutrality was described as "*sui generis*"[145], a position later upheld by the Maltese courts.[146] The definition of neutrality in the Constitution is that Malta is a neutral country adhering to a policy of non-alignment and refusing to participate in any military alliance.[147] While the bill introducing the Constitutional amendment was being debated in Parliament just prior to the 1987 election, the Socialist Prime Minister piloting it through the debate, stated most emphatically that 'the status of neutrality is based on the principle of non-alignment', while he clearly identified non-alignment with reference to the superpowers and their blocs.[148] The end of the Cold War and developments in Europe after the removal of the iron curtain have changed both the meaning and context of neutrality as well as of non-alignment. However, as already shown, the present government argues that in the present Mediterranean context neutrality still makes sense.

Neutrality does not mean that Malta renounces an active foreign policy. When Malta originally declared a policy of neutrality, in the years preceding and after 1979 when the British military bases in Malta were wound up, and years before Parliament entrenched its definition in the Constitution, the Socialist government (1971–87) still practised an active foreign policy, signing a neutrality treaty with Italy, a NATO country, another with the USSR, a friendship Treaty with Libya committing both sides to military co-operation and two secret treaties with North Korea aimed at strengthening 'the bonds of friendship and solidarity between the two peoples and armies.'[149]

Hence on the basis of past practice, Malta will find no difficulties in pursuing the same foreign policy lines as the rest of the member states of the EU. In security matters, Malta could adopt the same stance presently being taken by the EU neutrals, namely that the national defence will be organised by Malta, with the additional understanding that the military facilities will be used by foreign forces in the extreme situation of a threat or violation of its sovereignty. Further, Malta will be in a position to join in peace-keeping efforts of the Union. To prepare the way for this eventuality, Malta joined NATO's Partnership for Peace (PfP) in April 1995. The MLP declared that this move violated Malta's neutrality[150] and promised that it would pull Malta out of the partnership if elected,[151] which it did in October 1996.

One important factor to consider would be whether future evolution of the EU institutions would make it easier for Malta both to adhere to its neutrality and membership of the EU. This question is related to what institutional form a future CFSP will take following the changes which will take place in the EU's Intergovernmental Conference (IGC). The proposal which appears to be the most popular among member states at the IGC is that a mechanism will be devised whereby countries will have the option not to join a military action of the Union but will be obliged to show financial solidarity. This is the formula by which the European neutrals could most likely participate in the CFSP without difficulties. It is nearly the same status which Denmark has enjoyed since the decisions of the Edinburgh European Council.[152] The only difficulty is that as Calvocoressi has observed, even a limited obligation of this kind would impair the would-be neutral's claim to neutrality in the event of war, and to accept it implies, therefore, a certain view of European affairs – a view of Europe in which there will be no overriding case for claiming neutrality in order to preserve the state's integrity and existence.[153] This means that a formula is possible whereby a neutral country could be accommodated as a member of the EU, without obstructing the Union's overall effectiveness.

## THE FOREIGN POLICY PROCESS AND SOURCES

The biggest problem facing the micro-state in its foreign policy-making and formulation is the lack of adequate resources, both human and material. As far as the basic requirements for the gestation of an independent foreign policy are concerned, the micro-state is in a sense a larger state writ small. There is a minimum of resources which the micro-state has to siphon off from competing demands on them and which have to be devoted to the making and implementation of foreign policy. At the end of 1995, the Ministry of Foreign Affairs of Malta had an overall staff compliment of around 200, including the non-diplomatic grades.[154] Around 56 of these were stationed in Malta's embassies overseas, carrying out diplomatic duties with respect to 66 countries and 7 major international organisations, including the European Union, and a host of agencies of these organisations. If allowance is made for those Foreign Ministry employees in purely managerial grades, the number of staff actually dealing with diplomatic issues will be less than 200. The Ministry of Foreign Affairs accounts for around 1.42 per cent of the national budget. Desk officers at the Ministry are engaged in the task of gathering and processing information, much of which originates from the diplomatic missions abroad, effectively around 15 in all. Lack of adequate human resource makes the information gathering not only a Herculean task but leaves little for adequate analysis and compilation of reports. For this reason more sophisticated methods of information gathering and processing are precluded. Since independence, successive governments have roped in both non-Ministry local expertise, namely in the form of academics or practising lawyers as well as a number of other foreign advisers at times supplied by UN agencies or under the aegis of some bilateral arrangement. This intercourse between academics and practitioners, difficult as it is,[155] could perhaps be more extensive if the systematic study of international relations, diplomacy, international law and related subjects was well founded at the University. Indeed, the biggest problem lies in the fact that although Malta has a University whose origins go back to 1592, it now needs a department of international relations and what goes for international relations is in reality current affairs. The systematic study of European integration has largely come about as a result of the efforts of the University's European Documentation and Research Centre as well as the Department of European and Comparative Law of the Faculty of Law. An International Maritime Law Institute has been established by the International Maritime Organisation. International Law is one of the components of the Law degrees conferred by the University. In 1990, a Mediterranean Academy of Diplomatic Studies was established jointly by the Institute for International Affairs of Geneva and the University of Malta. The aim of this institution was the training of Maltese and foreign diplomats. From the pool of Maltese graduates, the government was in a position to recruit trained personnel for the foreign service. This was not however the only source of new recruits. Through appropriate agreements financed by the EU–Malta financial protocols, a number of Maltese nationals have annually concluded postgraduate study in the College of Europe in Bruges and the University of Sussex. The unit dealing with EU matters, the 'EU Directorate', is by far the best manned and equipped section of the Foreign Ministry.

In June 1992, the Ministry of Foreign Affairs started publishing a *Malta Review of Foreign Affairs* which in most editions so far has carried very few analytical articles. However, the reproduction of key official documents and speeches on foreign policy provide an excellent source of foreign policy material. The other sources are the usual stock of newspaper interviews, parliamentary debates and material published by the Department of Information. Despite the fact that external relations are so crucial for a micro-state, foreign policy does not occupy a very large share of the political debate in Malta. In common with many other countries, centre stage is occupied by domestic issues, unless these are linked to some aspect of foreign policy, such as the economic and legal reforms started as part of the Structured Dialogue with the EC which was connected with Malta's 1990 EU application. The presentation of the annual budget by the government, in which every vote of expenditure is debated in Parliament serves as another occasion for a public discussion of Malta's foreign policy in parliament which at times spills over into the media.

In the meantime the political parties' stand on various issues is known through their respective electoral manifestos and periodic statements. The formulation of policy is even more complicated. The history of foreign policy-making in Malta is very recent, having started in 1964 with Independence. So far it appears that normally either foreign policy is first formulated and moulded into a Party stand by a closed group of officials working close to the party leaders or at times by a small group of people close to the prime minister, not necessarily close to the Ministry of Foreign Affairs or any of the parties. Once in government, the minister responsible relies on these 'principles', thus formulated, as well as the advice of foreign office staff in the implementation of policies. Major decisions or initiatives normally require cabinet approval. Internal consensus on all issues within the parties and cabinet is not normal. The force of public opinion is a major constraining force on the freedom of the foreign policy-makers. When a major foreign policy issue emerges, the press normally takes keen interest. Domestic and foreign affairs are at times intertwined at others entangled. For instance, as an example of entanglement, Malta's status of neutrality was enshrined in the Constitution in a *quid pro quo* involving the right to majority rule, a purely domestic issue. As a case of intertwinement, membership of the EU and a number of domestic issues were debated as a continuum on many occasions during the 1990s.

Bureaucratic politics analysis is relevant, though given the small size of the bureaucratic set-up at the Ministry of Foreign Affairs, this does not promise to be a major avenue for the understanding of Malta's foreign policy. What is interesting is the inter-play between different ministries and the strain that these demands make on the meagre army of diplomats actually in the field. Ultimately, foreign policy-making remains closely connected with a few individuals and the careful analysis of parliamentary debates, press interviews, speeches and party statements remains the most promising avenue. For the past nine years the Ministry of Foreign Affairs has published an annual report and if this is continued it serves as another important source of information. There is however, little light thrown on internal decision-making structures or debate, no official or semi-official papers used in this internal appraisal of policies and their alternatives within the Ministry of Foreign Affairs ever become public. The cabinet debates and decisions constitute another dark area. From 1996, with the start of the new legislature after the election, a Parliamentary Committee dealing with Foreign Affairs started functioning. The establishment of the committee may engender the emergence of a greater consensus in foreign policy. Its workings may in the longer run augment the amount of information reaching the public, increase awareness, improve the debate generally and possibly lead to a deeper analysis of the problems. In the whole scheme of things what is missing is the proper structures on the academic side; if these existed the whole decision-making-analysis process of foreign policy in Malta could be more rounded.

**Editorial Note:**

Following an early election, on 5 September 1998, a Nationalist government was elected with a clear majority and its first decision was to reactivate the application for EU membership which was a prominent issue in the campaign.

## NOTES AND REFERENCES

1. Hans J. Morgenthau, *Politics among Nations: The Struggle for Power and for Peace* Alfred A.Knopf, New York, 6th edn, 1985, pp 115–81.

2. Ibid., pp 174 ff.
3. Kenneth N. Waltz, *Theory of International Politics*, Random House, New York, 1979, p.132.
4. Christopher Hill, Theories of Foreign Policy Making for Developing Countries, in C. Clapham (ed.), *Foreign Policy Making in Developing States*, 1977, pp 11 ff.
5. James N. Rosenau, *The Scientific Study of Foreign Policy*, revised edn, Frances Pinter, London, 1979, p. 371.
6. James N. Rosenau, *The Study of Global Interdependence: Essays on the Transnationalisation of World Affairs*, Frances Pinter London, 1980. See pp. 162 ff. Theorizing Across Systems: Linkage Politics Revisited.
7. James N. Rosenau, Pre-theories and Theories and Foreign Policies in Barry R.Farrel (ed), *Approaches to Comparative and International Politics*, Northwestern University Press, 1966; also, Rosenau J.N., *The Scientific Study of Foreign Policy*, pp. 115 ff. See especially Table 6.1, p.133. See also Neack Laura, Linking State Type with Foreign Policy Behaviour, in L. Neack, J.A.K. Hey and P.J. Haney (eds), *Foreign Policy Analysis: Continuity and Change in the Second Generation*, Prentice Hall, New York 1995, pp. 215 ff.
8. Ibid., p. 218. Further, Rummel's study quoted below seems to confirm this.
9. David Vital, *The Inequality of States: A Study in Small Power in International Relations*, Clarendon Press, Oxford, 1967, p.8.
10. Michael Handel, *Weak States in the International System*, Frank Cass, London, 1990, p.11.
11. Ibid., p.47.
12. Ibid., p.48.
13. Aron Waschkuhn, Small State Theory and Micro-states, in Frank R. Pfetsch, *International Relations and Pan-Europe: Theoretical Approaches and Empirical Findings*, Lit, 1993, pp. 149 ff.
14. A summary of the use of population size to classify states in the literature on small states is to be found in P. Hein, *The Study of Micro-states*, in J. Kaminarides L. Briguglio and H. N. Hoogendonk, *The Economic Development of Small States: Problems, Strategies and Policies*, Eburon, Delft, 1989, pp 10–11.
15. Waschkuhn (op. cit., p.151) quotes Liechtenstein (population 30,000), Which was refused membership of the League of Nations in 1920 and only joined the United Nations in 1990, and also refers to the UN Secretary General U Thant's call in 1965 for UN member states to examine more closely the criteria for the admission of new members.

16. Barry Buzan, *People, States and Fear: An Agenda For International Security Studies in the Post-Cold War Era*, Harvester Wheatsheaf 2nd edn, 1991, pp. 112 ff.
17. Handel, op.cit., pp. 52–53.
18. See Jeanne A. K. Hey, in L. Neack, J. A. K. Hey and P. J. Haney, op. cit., pp. 201 ff.
19. Valerie M. Hudson and Christopher S. Vore, Foreign Policy Analysis Yesterday, Today and Tomorrow, *Mershon International Studies Review*, No. 39, 1995, pp. 209–38.
20. Ibid., pp. 225–229.
21. Ibid., p. 222.
22. Maurice A. East, Size and Foreign Policy Behaviour: A Test of Two Models, *World Politics*, Vol 25, No 4, 1973, pp. 556 f.
23. Ibid., pp. 558–560.
24. Security here is seen both in the context of hard security as well as economic prosperity and stability. See Buzan in R. Stubbs and G.R.D. Underhill, *Political Economy and Changing Global Order*, MacMillan, 1994, pp.89 ff.
25. *Note verbale* circulated to all EU member states and published by the Department of Information, Malta, 23 November, 1996.
26. W. F. Stolper, E. R. Hellberg, and S. O. Calendar, *Economic Adaptation and Development In Malta*, Report prepared for the Government of Malta, United Nations Programme of Technical Assistance, 20 January, 1964, see point 59.
27. Ritchie Ovendale, "*British Defence Policy Since 1945*", Manchester University Press, 1994, p.14.
28. In a televised address to the nation, 1 October, 1971, Prime Minister Dom Mintoff referred to the Libyan aid but did not specify the amount. He again made reference to this steadfast support when addressing the Parliamentary Assembly of the Council of Europe on 28 September, 1978.
29. Recall that it was in the early 1970s that Sino-Soviet rivalry rose to a pitch, that Brezchnev contemplated a nuclear attack on China, and that he actually solicited American acquiescence on the matter.
30. Statement by the Prime Minister of Malta, Dr E. Fenech Adami to the Ambassadors and Other representatives accredited to Malta, 9 June, 1987, Department of Information (DOI), p. 4.
31. Foreign Minister Dr George Vella during a visit to Bonn as reported in *The Times (of Malta)*, 13 March, 1997.
32. John C. Grech in *Threads of Dependence*, University of Malta Press, 1978, quotes the report of Sir George E. Schuster,

*Interim Report on the Financial and Economic Structure of the Maltese Islands*, London, HMSO, 1950, p. xviii, as expressing clearly that Malta's economic viability depended on its ability to export.

33. A discussion of the alternative scenarios of a trade pact between Malta and either EFTA or the EEC is discussed in the Final Report to the Government of Malta by F. K. Liebech, United Nations Trade Policy Adviser, December 1965, see points 23–4, pp 11–12 . Arguments in favour of an open economic policy are also to be found in the report to the Malta government by Stolper *et al.*, op.cit, point 19, p. 8. Also, Liebich F.K., *Revision of the Customs and Trade Policy of Malta*, Valletta, August 1964, points 30–34, pp. 12–13.
34. The development of relations between the EEC and Malta is treated in Pace, Roderick, *Malta u l-Komunita Ewropea: 1961 – 1990*, PIN Publications, 1990 (Maltese); also R.Pace, 'Maltese EC Application: Some Economic and Political Considerations', *Bank of Valletta Review*, no. 1, Spring 1990, p. 1 ff.
35. The EEC–Malta Association Agreement was published in the *Official Journal of the EC*, L61, 14 March, 1971.
36. The Prime Minister, Dr George Borg Olivier in a speech during the signing of the Association Agreement in Valletta in December 1970, said that Malta saw the agreement as a step towards eventual membership. See text of the speech published in '*Malta Today*', vol.V, no. 12, 1970.
37. Richard Pomfret, *Mediterranean Policy of the European Community: A Study of Discrimination in Trade*, St Martin's Press, New York, 1986, pp. 63–74.
38. *Economic Survey 1996*, Economic Planning Division, Ministry of Finance, Malta, 1996.
39. The 'Special Relationship' was outlined in a speech by Prime Minister Dom Mintoff to the EC–Malta Council of Association which met in Brussels in June 1981. The speech was published by the DOI in June 1981 as a supplement to the official publication, '*Il-Gzejjer*'.
40. Details can be found in R. Pace, 'Assessing Malta's Bid to Join the European Union: The Case in Favour', *Discussion Papers in European and International Social Science Research*, no. 54, University of Reading , July 1995.
41. See the statement by the Prime Minister in Parliament, '*Ir-Review*', 11 October 1969, Department of Information, Valletta.
42. Communiqué reported in *The Times (of Malta)*, 27 April 1972, p.20.

43. This can be deduced from the publication of the official documents exchanged between Malta and Libya on the issue. See *Official Documents About the Malta/Libya Dispute on the Dividing Line of the Continental Shelf*, DOI, Malta, 8 September 1980.
44. There are two versions of what happened on the high seas. Mintoff accused Libya of using force against Malta. Libya claimed that no force was used and that its warships sheltered a Maltese patrol boat sheltering in the area from the rough weather. See E. Saliba, *Malta and the Security of the Mediterranean Region* in a publication by the same title, International Security Council, January 1987, pp. 72–3.
45. Extract from a Malta Government statement issued on 27 August 1980.
46. Agreement between Libya and Malta was concluded on 23 May 1976 through a so called special agreement signed by Maltese Minister W. Abela and Taha Sherif Ben Amer, Libyan Minister of State for Revolutionary Command Council Affairs. According to Article IV, the agreement was to enter into force on the date of exchange of the instruments of ratification by the two governments and was to be notified jointly to the Registrar of the Court.
47. Procès-verbal of the exchange of instruments of ratification of the special agreement between the Socialist People's Libyan Arab Jamahiriya and the Republic of Malta for the Submission to the International Court of Justice of Difference, signed in Valletta on 20 March 1982.
48. Protocol on 'Co-operation on Security' attached to the 'Treaty of Friendship and Co-operation between the Government of the Republic of Malta and the Socialist People's Arab Jamaharija', signed in Valletta on 19 November 1984 by Prime Minister Dom Mintoff and Muammar El Gaddaffi.
49. Article 6 of the Treaty.
50. A. Smith, 'Malta: From NATO Base to Libyan Outpost' in *Malta and the Security of the Mediterranean Region*, International Security Council, Milan, 23–25 November 1986, pp. 31–2.
51. The Prime Minister (Mintoff's successor) Carmel Mifsud Bonnici said this in an interview with the *New York Times*, Reuters Service, New York, 6 August 1986.
52. According to the *New York Times*, 29 October 1988, quoting Gaddaffi speaking to Italian journalists, 'It was the Prime Minister of Malta who telephoned us to advise us they were preparing an attack on us and that America wanted my head that night. The Italians told Bonnici (the Prime Minister)'.

53. Statement by the Prime Minister, Dr E. Fenech Adami to ambassadors and other representatives accredited to Malta, 9 June 1987, DOI, Malta.
54. 'Malta Ends Intelligence Agreement with Libya', *Jane's Defence Weekly*, vol. 10, no. 12, 24 September 1988.
55. The new Treaty was initialled in Valletta in November 1989 and signed in Valletta on 19 February 1990 by the Maltese Prime Minister Dr Eddie Fenech Adami (then also responsible for Foreign Affairs) and Eng. Jaddalah Azzuz Al-Talhi, Secretary of the People's Committee of the People's Bureau for Foreign Liaison and International Co-operation of the Socialist People's Arab Jamaharija.
56. *In-Nazzjon Taghna*, 28 October 1988.
57. In a statement by the Foreign Minister of Malta, Dr Vincent Tabone, Malta offered to restore normal relations between the two sides. See In-Nazzjon, 5 January 1989.
58. The leader of the Opposition, Dr Carmelo Mifsud Bonnici, reported in *It-Torca*, 15 January 1989.
59. Resolution 883 (1993) adopted by the Security Council of the UN at its 3212th meeting on 11 November 1993.
60. Leader of the Opposition, Dr Alfred Sant, reported in *The Times (of Malta)*, 1 December 1993.
61. Malta and Libya have established a mixed commission which meets every six months and which directs trade and other commercial relations between the two sides.
62. See *The Malta Independent*, 31 March 1996.
63. See report in *The Times (of Malta)*, 1 November 1995; also, 'Malta–Libya Relations Pass Through Anxious Moments', *Malta Business Weekly*, no. 54, 2–8 November 1995.
64. Report (fourth) of the Ministry of Foreign Affairs, '*Ir-Raba Sena Tat-Tieni Legislatura Tal-Gvern: Rapport ta' Hidma 1995–96*', Ministry of Foreign Affairs, Malta, 1996, p. 51.
65. Following talks in Tripoli, the Maltese Foreign Minister Guido de Marco said that relations between Libya and Malta had returned back to normal. See *The Times (of Malta)*, 30 December 1995.
66. Israel's ambassador to Malta, Yehuda Millo, was reported to have claimed that the episode had not affected negatively relations between Israel and Malta, *The Times (of Malta)* 9 February 1996.
67. *The Malta Independent* of 24 December 1995 claimed that soon after the Shaqiqi assassination, the Libyan leader Mummar Gaddaffi had urged Malta to review its relations with other

states in the region if continued friendly ties with Libya were desirable.

68. Reported by *The Malta Business Weekly*, issue of 17–23 August 1995.
69. Quoted in *The Malta Independent*, 20 August 1995.
70. *The Malta Independent*, 20 August 1995.
71. *The Times (of Malta),* 1 November 1996.
72. Stolper *et al.*, op.cit. (note 26), point 67 ff., pp. 25 ff.
73. The issue was discussed by Maltese Minister G. Felice with Italian Foreign Minister Emilio Colombo and was reported in '*Ir-Review*', of 20 June 1962.
74. Joseph Bugeja, '*Italo-Maltese Relations 1880–1980*', unpublished MA thesis, Mediterranean Academy of Diplomatic Studies, University of Malta, Academic Year 1994–5.
75. The agreements between the government of Malta and the government of the Italian Republic during this period were the following: - Concerning Scientific and Technical Co-operation, 28 July 1967 – Treaty Series No 40 - Cultural Agreement, 28 July 1967 – Treaty Series No 43, Relating to Economic Co-operation and Protection of Investments, 28 July 1967 – Treaty Series No 45, Trade Agreement, 28 July 1967 – Treaty Series No 51, Financial Agreement, 6 July 1968 – Treaty Series No 101 (3000 million Italian Lire)
76. The Agreement was signed on 23 May 1972.
77. E. Saliba, in op cit., p.72.
78. Protocol I (signed 15/9/1980) 60 million US dollars, Protocol II (signed 20/11/1986) 180 billion Italian lire, Protocol III (signed 5/11/1990) 250 billion Italian lire, Protocol IV (signed 28/3/1994) 250 billion Italian lire.
79. The main issues raised in the debate are summarised by Serracino-Inglott, Peter, 'Maltese Neutrality at the Italian Parliament', *AZAD Perspektiv*, no. 15, April–June 1981, pp. 5ff.
80. The long political crisis started in Italy after magistrates started investigating bribes in politics. This in turn led to the departure from power of many prominent Italian politicians and to the 'end of the first republic'.
81. 'New Leaf in Relations with Tunisia', *The Sunday Times (of Malta)*, 14 January 1996.
82. Malta's land territory is 316km square. The waters and underlying continental shelf which definitely fall under present Maltese jurisdiction are nearly 100 times its land area (approximately 30,000 km square) and could extend up to 200 times the land area if Malta's claims of further jurisdiction

are accepted. *Malta Structure Plan*, vol. 2, Section R, pp. 5, DOI, Malta, 1990.

83. 1 December 1964.
84. Address by the Prime Minister, G. Borg Olivier to the United Nations General Assembly, 6 October 1967, DOI, Malta, 1967.
85. The speech is reproduced in 'The Common Heritage: Selected Papers on Oceans and World Order 1967–74', *IOI Occasional Paper* no. 3, Malta University Press, 1975, pp. 1–42.
86. A very interesting account on how Pardo allegedly made the proposal is the one by Victor Gauci in 'Towards a Second Generation United Nations', Special Issue, *Malta Review of Foreign Affairs,* Ministry of Foreign Affairs, October 1993.
87. Statement to the Plenary Session of the UN General Assembly by Malta's Ambassador to the UN, A.Pardo, reproduced in *IOI Occasional Paper* No. 3, op.cit., pp. 43 ff.
88. See details in 'International Seabed Authority: Malta's Former Ambassador Presents His Case', *The Times (of Malta)*, 28 July 1976, pp. 3–4; 'International Seabed Authority: Government's Stand', reply by the Director of Information, *The Times (of Malta)*, 31 July 1976; reply by Pardo, *The Times (of Malta)*, 9 September 1976.
89. P. Serracino-Inglott, op. cit.
90. Address by Eddie Fenech Adami, Prime Minister of Malta to the 44th Session of the General Assembly, 28 September 1989, DOI, Malta, p.4.
91. Statement by Dr Vincent Tabone, Minister of Foreign Affairs of Malta at the Second Meeting of Foreign Ministers of the Mediterranean Members of the Non-Aligned Movement, Brioni, 3–4 June 1987; DOI, Malta, 1987.
92. This was stressed at the NAM Foreign Ministers Meeting which took place in Nicosia in September 1989. See 'It-Tieni Sena Tal-Gvern: Ministeru ta' l-Affarijiet Barranin', *Report for the Year May 1988–May 1989 of the Ministry of Foreign Affairs*, Ministry of Foreign Affairs, 13 May 1989.
93. The Commission's Opinion on Malta's application makes it clear that 'the principle of neutrality and of Malta's non–aligned status set out in the Maltese constitution raise the problem of their compatibility with Title V of the Maastricht Treaty and could lead to difficulties in the area of 'Joint action' and future co-operation on defence', *Com (93) 312 final*, 30 June 1993, point 19, p.10.
94. Foreign policy document, Malta Labour Party (undated document circa 1996) pp. 19–20.

95. Malta Labour Party Electoral Manifesto 1996 election *Ic-Cittadin L-Ewwel: Vizjoni Ghal Malta Moderna*, p.2.
96. This point is discussed in Pace, R., 'Assessing Malta's Bid to Join the European Union: The Case in Favour', *Discussion Papers in European and International Social Sciences Research*, no. 54, July 1995, pp. 21–8.
97. The Constitution of Malta, DOI, 1992, Chapter 1.
98. Organisation for Security and Co-operation in Europe.
99. Statutory Report of the Committee of Ministers of the Council of Europe, Presentation by Dom Mintoff, Prime Minister of the Republic of Malta and Minister of Foreign Affairs, 28 September 1978, DOI, Malta, 1978.
100. In a straight two-party context and in an electoral system based on proportional representation, the Malta Labour Party with around 49 per cent of the valid votes cast obtained more Parliamentary seats than the Nationalist Party with nearly 51 percent of the vote.
101. A number of documents can be consulted on these issues including: (a) Resolution in the European Parliament 'On the Violation of the Freedom of the Press and Private Property in Malta', *OJ C 302*, 3 December 1979, p. 6; (b) the Den Heuvel Report, *Doc I-368/83* of the European Parliament; (c) CoE 'Report on The Foreign Interference Act and Other Issues Affecting the Relations between Malta and the Council of Europe', Rapporteur Mr Schwarz, *A Doc 5389*, April 1985.
102. Act XIV of 1987, *Supplement to the Government Gazzette*, no. 14,842, 19 August 1987, DOI, Malta.
103. Address by the Prime Minister, E.Fenech Adami to the Parliamentary Assembly of the Council of Europe, 28 September 1990, DOI, Malta.
104. John J. Maresca, *To Helsinki: The Conference on Security and Cooperation in Europe, 1973–75*, Duke University Press, 1987, pp. 189 ff.
105. Ibid., pp. 191–2.
106. The meeting was held between 13 February and 26 March 1979.
107. 'The Meeting of Experts concluded that for many of these sectors (fields of cooperation identified in the meeting) ECE, UNESCO and UNEP provided a framework for investigating and developing further cooperation among the participating states and the non-participating Mediterranean states...', *Report of the Meeting of Experts* , Valletta, 2 March 1979.
108. A. Bloed, '*From Helsinki to Vienna: Basic Documents of the Helsinki Process*', Dordrecht, Nijhoff, 1990, p.16.

109. *Agreement on the Neutrality of Malta, Malta–USSR*, 8 October 1981, Ministry of Foreign Affairs, page 8 of the published text.
110. E. Fenech, Adami, 'Conflict and Unity in the Mediterranean: A Sea-Centred Functionalist Approach', speech at the University of Toronto, 25 May 1983.
111. David Mitrany, 'A Working Peace System' (1943), in P. Taylor and A.J.R. Groom (eds), *The Functional Theory of Politics*, London 1975; 'The Prospect of Integration: Federal or Functional', *Journal of Common Market Studies*, vol. 4, 1965, pp. 119ff.
112. Convention for the Protection of the Mediterranean Sea Against Pollution and its Related Protocols, which came into force on 12 February 1978. The texts were published by the United Nations Environment Programme (UNEP) in 1982.
113. The Speech by the Foreign Minister in which the idea of a Council of the Mediterranean was outlined was published in *The Sunday Times (of Malta)* of 22 November 1992.
114. *L-Ewwel Sena Tat-Tieni Legislatura tal-Gvern*, Report by the Foreign Minister covering the 12-month period beginning in February 1992. Some confusion exists as to how the original idea came about. The Department of Information claimed that the idea had actually been put forward at a meeting in Algiers in 1990 by the Maltese foreign Minister. The Labour Party also claimed credit for it.
115. E. Fenech Adami, Prime Minister of Malta, Mediterranean Security: A Maltese Perspective', *Mediterranean Quarterly*, vol. 6, no. 1, 1995, pp. 1–9.
116. Ibid., page 7.
117. Ibid., pp. 8–9.
118. Malta Labour Party, '*Foreign Policy Document*' (undated but published in 1996), pp. 9–10.
119. *Malta Review of Foreign Affairs*, No 9, May 1996, Ministry of Foreign Affairs, p. 46.
120. E. Fenech Adami, Malta in the European Union, Brussels, 3 February 1994.
121. *MLP Electoral Manifesto*, 1996, p.2.
122. Foreign Policy Document, op.cit., p.7.
123. As Dr Sant was reported in *The Times (of Malta)* of 15 October 1996.
124. Press statement by the Nationalist Party, *The Times (of Malta)* 16 October 1996, p.12, in which the Party described Labour's proposal thus: 'the MLP was the same party which allowed the green flags of Libya to replace those of the United Kingdom after the end of the military bases in March 1979'.

125. *The Times (of Malta)*, 12 March 1997.
126. *The Times (of Malta)*, 5 April 1997.
127. *The Times (of Malta)*, 5 April 1997.
128. *The Times of Malta*, 8 April 1997.
129. *The Times (of Malta)*, 25 April 1997.
130. During the recent election campaign, Prime Minister Alfred Sant was reported to have said that 'I think that Malta will again be seen as a fortress if it joins the EU, and this will be detrimental to stability in the Central Mediterranean', *The Times (of Malta)*, 15 October 1996.
131. *MLP Electoral Manifesto*, 1996, p. 2.
132. Ibid., p. 35.
133. *Foreign Policy Document*, 1996, p.17: 'it is also in the best interest of our national security to safeguard and strengthen the neutrality and Non-Aligned status of our country'.
134. Ibid., p.17.
135. Ibid., pp. 16–21.
136. Ibid., p. 20.
137. Quoted by J.F. Freymond, in 'Swiss Neutrality and The Future of Switzerland in a Changing Europe', in S. Harden (ed.) *Neutral States and The European Community*, Brassey's, 1994, p.3.
138. Ibid., p.3.
139. David E. Bohn, 'Neutrality – Switzerland's Policy Dilemma: Options in The New Europe', *Orbis*, vol. 21, no 2, 1977, pp. 335 ff.
140. Ibid., p. 339.
141. *Foreign Policy Document*, 1996, p.3.
142. A. Vasconcelos, 'Security in Northern Africa: Ambiguity and Reality – General Framework and Concepts' in A. Vasconcelos and F. Faria (eds), Security in Northern Africa: Ambiguity and Reality, Chaillot Papers no. 25, Institute for Security Studies, WEU, Paris, September 1996, p.14.
143. Commission's Opinion, *Com (93) 312 final*, Brussels, 30 June 1993, point 19, p.10.
144. Guido de Marco, Foreign Minister, Speech to the WEU Assembly, *The Sunday Times (of Malta)*, June 19th, 1994.
145. Statement by the Prime Minister Dr E. Fenech Adami to the Ambassadors and other Representatives accredited to Malta, 9 June 1987, p.5.
146. Case 'Saviour Balzan *noe et versus* The Honourable Prime Minister', 23 June 1988.
147. Chapter 1, Article 1 of the Constitution , section (3) (abridged) 'Malta is a neutral state . . . adhering to a policy of non-alignment and refusing to participate in any military alliance. Such

a status will, in particular entail that: (a) no foreign military base will be permitted on Maltese territory; (b) no military facilities in Malta will be allowed to be used by any foreign forces except at the request of the Government of Malta and only in the following cases: (i) in exercise of the inherent right of self-defence...or in pursuance of measures...decided by the Security Council of the UN, or (ii) whenever there exists a threat to the sovereignty, independence, neutrality...of the Republic of Malta...' The shipyards are denied to the military vessels of the two superpowers.

148. *Debates of the House of Representatives*, entry for 30 December 1986, p. 93.
149. The first agreement was signed in Malta in March 1982 while the second Treaty was signed in July of the same year. Both sides were to observe strict secrecy on the accords. The treaties were published after the change in government in 1987.
150. Press Communiqué issued by the MLP and published in the Maltese language daily *L-Orizzont*, 9 December 1994.
151. Resolution approved by a general conference of the MLP on 1 April 1995 (text).
152. Section of the Conclusions of the Presidency of the Edinburgh European Council of 11–12 December, 1992, by which Denmark is permitted not to participate in the elaboration and implementation of decisions and actions of the Union which have defence implications, without (Denmark) preventing the adoption of such Council decisions.
153. Peter Calvocoressi, 'Neutrality Now' in S. Harden, *Neutral States and the European Community*, Brassey's, London, 1994, p.156.
154. *Estimates 1996*, Ministry of Finance, 1995, pp. 65 ff.
155. See for example the problems discussed in M. Girard, E. Eolf-Dieter and K. Webb (eds), *Theory and Practice in Foreign Policy Making: National Perspectives on Academics and Professionals in International Relations*, Pinter, London, 1994.

# 9 Turkey and the Mediterranean

Kemal Kirişçi

## INTRODUCTION

During the Cold War, Turkey's foreign policy was mostly a function of its geographical location as a neighbour of the Soviet Union. Soviet territorial demands over Turkey in 1945 and the fear of Communism were critical in compelling Turkey to seek the support of the West for its defence. Once it joined NATO and other western inter-governmental organisations Turkey's foreign policy became basically dominated by its relations with Western Europe and the United States. The Mediterranean would have never attracted the attention of Turkish foreign policy-makers other than in the context of NATO's military strategies. Conflicts with Greece and the Cyprus problem since the early 1960s were additional issues that were important to Turkish foreign policy. However, it would be difficult to argue that these issues increased the relevance of the Mediterranean as a region to Turkish foreign policy. Instead these were basically seen as bilateral issues.

In the 1950s there was a brief period when Turkey was actively involved in Western efforts to prevent Soviet influence from expanding into the Middle East but this involvement did not last very long: once pan-Arab and Baathist regimes in Egypt, Iraq and Syria came to power, Turkish foreign policy became characterised by non-involvement and non-interference in the politics of the Middle East. During this period Turkey maintained balanced relations with Arab countries and Israel with the exception that from the mid-1970s Turkey tried to expand its economic relations with oil-rich Arab countries. During this period Arab countries and Israel were perceived as a part of a region called the Middle East and North Africa rather than the Eastern Mediterranean.

The end of the Cold War and the disintegration of the Soviet Union has brought about major changes to Turkish foreign policy. In terms of regional politics Turkey has stopped being simply an outpost of the West. Instead, Turkish foreign policy began to feel the demands of belonging to numerous geographical and political regions. The collapse of Communism in the Balkans and the disintegration of Yugoslavia drew Turkey into the politics of the region. The rise of ethnic violence and repression of Turkish and Muslim minorities in the Balkans created domestic pressure within Turkey to take an interest in the Balkans. Furthermore, the danger of the war in the former Yugoslavia engulfing the rest of the Balkans was an additional concern directly affecting Turkish security and ground links with West European markets for Turkish products.

Similar observations can also be made about the Caucasus and Central Asia. In the Caucasus, the emergence of Armenia, Azerbaijan and Georgia as independent states and the fact that, including in the Russian Federation, there are ethnic groups with strong links and affinities to Turkey, have created a complex foreign policy environment. Furthermore, economic, environmental and political factors have fuelled fierce competition over the routing of pipelines that will carry Caucasian and Central Asian oil to western world markets. At the same time, the emergence of independent Central Asian republics with ethnic and cultural links to Turkey opened up new economic, social and political opportunities that no foreign policy-maker had envisaged during the Cold War. Yet, in respect to Central Asia decision makers had not only to cope with the euphoria of the 'discovery' of Central Asia and the unrealistic expectation of forming a community of Turkic states but also manage its relations with Russia, the traditionally influential power in the region.

The Middle East is another area where Turkish policy has changed dramatically. The change occurred after the Iraqi invasion of Kuwait. The decision to support the efforts of the United Nations to expel Iraq from Kuwait and to allow the creation of a safe haven in northern Iraq for more than one and a half million Kurdish refugees who had fled to Iran and Turkey undermined the traditional Turkish foreign policy principles of non-interference and non-involvement in Arab affairs. Furthermore, the beginning of first the Arab–Israeli and then the Palestinian–Israeli Peace Process led Turkey to expand its economic and political relations with Israel. These developments have also coincided with a period during which Turkish–Syrian and Turkish–Iranian relations have been complicated by the Kurdish problem and the trans-border activities of the separatist Kurdish rebel group, the Kurdistan Workers' Party (PKK).

One last area that has seen its importance increase is the Black Sea region. During the Cold War the Black Sea was simply a sea that separated Turkey from the Soviet bloc countries. There was very little interaction that took place across the Black Sea. However, this situation changed dramatically with the end of the Cold War. Ex-Soviet bloc citizens poured into Turkish cities for trading purposes. This played a critical role in the Turkish government's initiative to start up the Black Sea Economic Cooperation to encourage commerce and greater cooperation among littoral countries. Compared to the Cold War years, it is particularly Turkey's trade with Russia, and to some extent the Ukraine, that has increased substantially. Furthermore, many Turkish companies of all sizes have become involved in economic activities in practically all of the Black Sea countries. It is possible to argue that these developments have led to a discernible Turkish foreign policy towards the region covering a wide range of issues ranging from environmental cooperation to conflict management.

The end of the Cold War has accentuated the role of geography in a manner very different than was the case at the end of the Second World War. Today, Turkey finds its foreign policy to be very much a function of the fact that it is geopolitically part of the Balkans, the Black Sea region, the Caucasus and the Middle East and that politically it is associated with the West and also with Central Asia. However, in spite of Turkey having a long coast on the Mediterranean Sea and in spite of the Mediterranean region acquiring growing political and economic importance for many countries, the Mediterranean still remains a secondary consideration for Turkish foreign policy. For many Turkish decision makers, Chipman's observation before the end of the Cold War, that the Mediterranean is a 'geographical entity' and that this 'does not make it either a political or strategic whole',[1] is still valid today even if this may well change in the near future.

The thesis of this chapter will be based on the argument that current Turkish foreign policy is very much dominated by a wide range of economic, political and security issues that mostly link Turkey, in the minds of decision-makers, to geographical regions other than the Mediterranean. However, it is possible to argue that these regions have become interrelated and that the interdependence between issues make the boundaries of these regions overlap. There are a number of issues that are common in terms of their effect on regional security and stability. Economic disparities, ethnic rivalries, terrorism, migration, drug trafficking, environmental problems, the threat of ballistic missiles, are common concerns to both ends of the Mediterranean as much as to the other regions that Turkey geographically borders. Hence, in respect to addressing these transnational and in many ways transregional issues Turkey is in a critical location, geographically and politically. Therefore, it should be possible to agree with Williams' remark that Turkey is 'in many ways the hinge between the Community and the Eastern Mediterranean/Middle East and is also capable of playing an influential regional role with regard to southern republics of the former Soviet Union'.[2]

## DOMESTIC SOURCES OF FOREIGN POLICY AND FOREIGN POLICY ANALYSIS (FPA)

Clarke notes the importance of models in analysing foreign policy behaviour.[3] Hudson provides an excellent review of the literature on foreign policy analysis and especially recent conceptual developments.[4] However, by and large, neither the concept of foreign policy analysis nor the wide literature on the topic appear to have affected analysis of Turkish foreign policy in any significant manner. Literature that focuses on Turkish foreign policy has generally been atheoretical or at best made an implicit use of Allison's 'Rational Actor Model'. Allison had observed that most analysts rely on this Model in explaining the behaviour of national governments.[5] Today, with a few exceptions, his remarks are still valid in respect to Turkish foreign policy analysis in the 1990s.

Özcan[6] and Tanör[7] offer a systematic analysis of the role of institutions such as the military and the foreign ministry in

Turkish foreign policy-making. Aykan[8], Kut[9], Önis[10] and Winrow[11], include in their analysis the role of domestic politics. Saribay[12] and Özel[13] refer to the role of cognitive and psychological factors. Vali's *Bridge across the Bosporus* can still be considered to be a classic work on Turkish foreign policy that provides a comprehensive coverage of domestic sources of foreign policy ranging from ideologies, political parties, public opinion and government institutions.[14] However, with the exception of Saribay, the others do not make any reference to existing theoretical literature in their analysis.

The reasons for this are numerous. Firstly, academics outside Turkey who have taken interest in Turkish foreign policy have tended to come from a country specialist background with little grounding in FPA. It is not theoretical concerns that have been the driving factor behind their analysis. Describing Turkish foreign policy has usually taken precedence over theorising about Turkish foreign policy. Secondly, access to relevant information and data suitable to test theoretically derived hypotheses on foreign policy is still very difficult. The tradition of secrecy in the Turkish bureaucracy, especially in the military and the foreign policy establishment, remains very strong. Naturally, this makes it almost impossible to examine the role of what Clarke calls 'complex processes of group dynamics' and 'the information processing characteristics'[15] of foreign policy decision-making systems in Turkey.

Last but not least, as a discipline, international relations, as opposed to political science, is very recent in Turkey. The number of Turkish academics exposed to the theoretical literature in international relations and accustomed to think theoretically, in the sense that Rosenau has described it, is very limited.[16] Instead, the analysis of foreign policy and international relations has been considered for a long time to be the preserve of retired diplomats and high-ranking military officers. It is not surprising that until recently a large proportion of the literature in Turkey on foreign affairs was mostly written by such people. Most of them did not have any formal education in the discipline of international relations. Hence, their analysis have mostly been the product of experience and/or of institutional socialisation rather than theoretical thinking.

A number of general conceptual observations can be made in respect to domestic aspects of foreign policy-making in Turkey. Technically speaking, Vali's remark that 'in Turkey, as in other parliamentary regimes, the Council of Ministers, headed by the prime minister and advised by the minister of foreign affairs, is the chief decision-making organ in foreign affairs' is correct.[17] However, two institutions continue to have a significant influence on foreign policy-making: the Ministry of Foreign Affairs and the military establishment. To the institutional factors it is also important to add the cognitive aspects of Turkish foreign policy-making as well as the role of political parties and public opinion.

In Turkey, the Ministry of Foreign Affairs has traditionally been held in high esteem. The high quality of the personnel that serve in the Ministry and partly the general belief that the Ministry is bipartisan play an important role in this. The Ministry's general approach to foreign policy can be characterised by its pragmatism, conservatism and pro-western stance. The pragmatism of the Ministry is best demonstrated in the case of Turkey's relations with Iran. In spite of the generally shared belief that Iran supports Islamic fundamentalist groups as well as terrorism in Turkey, the Ministry has preferred to play down conflicts with Iran. The same can be said in respect to Turkey's relations with Syria. Ideological considerations and public opinion are often downplayed in favour of approaches that avoid confrontation.

At the height of the Armenian–Azeri war when the then President of Turkey, Özal, had suggested 'dropping a few bombs on Armenia and sending a brigade of soldiers would be enough to stop Armenian aggression'[18], the Ministry strongly counselled against the idea. This led the Prime Minister, Demirel, to remark that 'Turkey would not act alone militarily' and argue that 'foreign policy decisions cannot go along with street level excitement'.[19] Instead, the government chose to work through multilateral fora. At the United Nations Security Council it played a critical role in seeing through the adoption of resolutions supporting the territorial integrity of Azerbaijan. In a similar manner it supported the work of the Minsk Group within the Organisation of Security and Cooperation in Europe as a vehicle for mediation in the conflict between the two countries.[20]

A similar observation can also be made in respect to the Imea/Kardak crisis between Greece and Turkey. At the height of the crisis when the Greek military landed troops on one of the two islets that constitute Imea/Kardak, the Ministry strongly advised against the Turkish General Staff's plan to use force which was supported by the then Turkish Prime Minister, Çiller. The position taken by the Ministry led instead to a compromise plan which culminated in the landing of Turkish troops on an adjacent, empty, islet. This may have well been a critical move that prevented the crisis from escalating to a fully fledged armed confrontation between the two countries, enabling the US mediation effort to successfully diffuse the crisis early in February 1996.

There have also been occasions when the Ministry's conservatism and reluctance to take risks have attracted criticisms. In July 1993 during his tour of the Middle East, the then Minister of Foreign Affairs, Çetin cancelled what would have been the first visit of a Turkish Minister to Israel when the Israelis mounted military attacks on southern Lebanon on the eve of his visit. The visit had been prepared months in advance and was hailed as a sign of a turning point in Turkey's policy towards the Middle East. The fact that the cancellation occurred shortly before the Israeli–PLO Declaration was announced in September 1993 led to bitter criticism of the Ministry of Foreign Affairs for being out of touch with developments. It was argued that the cancellation of the visit had dealt a serious blow to the possibility of Turkey playing a role in the Middle East Peace Process.

In terms of the internal dynamics within the Ministry one consequence of the reluctance to take risks and the institutionalised preference for maintaining the status quo has been a lack of innovation on critical issues affecting Greek–Turkish relations. Ministry officials have been reluctant to call for policy changes because of the fear of failure. Here it is important to note the influence of what Janis defined as 'groupthink' on members of the Ministry.[21] The pressure to be part of an existing consensus and not challenge it is certainly an important obstacle in the way of innovative thinking in Turkish foreign policy-making. This seems to suggest that on most foreign policy issues the Ministry of Foreign Affairs will remain a conservative and status quo oriented institution. As MacKenzie points out, high-ranking bureaucrats of the Ministry 'rarely see it as their job to recommend bold initiatives or to formulate new approaches to intractable problems'.[22] Lastly, it should also be re-emphasised that the Ministry will remain an important pro-Western force in Turkish foreign policy-making, barring any significant domestic political changes.

The influence of the Ministry in foreign policy-making revealed itself in a very conspicuous manner after the formation of the new coalition government between the Islamic Welfare Party (Refah) and the conservative True Path (DYP) parties in June 1996. The Ministry, together with the military, played a very important role in persuading the new Prime Minister and leader of Refah, Erbakan, to reconsider his opposition to Operation Provide Comfort, the US-led military force based in southern Turkey to enforce the no-fly zone over northern Iraq. Erbakan was left in the embarassing position of having to persuade members of his own party in the parliament to accept a U-turn and vote in favour of the extension of Operation Provide Comfort's mandate for another term. Also, Erbakan found himself having to accept the continuation of close relations with Israel including the implementation of the military cooperation agreement signed between the two countries in February 1996. This was in spite of his objections to close relations with Israel and previous promises that once in power he would scrap the agreement.

The Foreign Ministry also opposed parts of Erbakan's Africa tour to Egypt, Libya and Nigeria in October 1996. The Ministry had advised Erbakan against a poorly prepared visit at short notice to Egypt and objected to his plans to visit Libya and Sudan on the grounds that both countries were widely recognised as countries supporting terrorism. Although Sudan was eventually excluded from his itinerary, Erbakan did go ahead with his visit to Libya against the advice of the Ministry. Subsequently, the Libyan leader's statements to Erbakan that, like all nations, Kurds too deserve to have a state of their own and that Turkey should not prevent this, left Erbakan in an extremely difficult situation. The Ministry recalled the Turkish Ambassador to Libya while Erbakan was left facing a massive barrage of public criticism in Turkey. In the parliament the government barely survived a motion of censure. After this experience Erbakan seems to have preferred to play a lower profile on foreign policy issues, leaving the field to his coalition partner Çiller, who held the position of Deputy Prime Minister as well as that of Minister of Foreign Affairs.

The military, especially on issues that are considered to have a direct impact on the security of Turkey, enjoy a powerful role in Turkish foreign policy. The influence of the military is usually exercised via the National Security Council (NSC). The NSC was originally established in 1960 to improve coordination between the government and the military as well as to advise the government on defence and security issues. The NSC is composed of the President, the Prime Minister, some members of the cabinet, the Chief of General Staff and the serving force commanders.[23] Technically, its decisions are supposed to be of a recommendatory nature. However, by and large the government tends to go along with its decisions.[24] The NSC has been influential particularly on policies concerning Operation Provide Comfort, the situation in northern Iraq, relations with Turkey's neighbours and the Armenian–Azeri conflict. The military has also been a strong advocate of close relations with Israel and operates as a powerful lobby in this area.

Clarke notes that the political culture of a country and the socialisation of leaders and officials have an important impact on their perception of the international environment and in turn on the decision-making process.[25] Snyder et al. by introducing the concept of the 'definition of the situation' as viewed by the decision-makers were among the first scholars to point out the importance of perceptions in foreign policy-making.[26] Jervis demonstrated in a convincing manner how decision-makers filter information from the outside world through their belief systems.[27] In this respect the military plays an indirect but fundamental role in the formation of Turkey's belief systems. All men in Turkey serve in the army and most top bureaucrats attend a training programme at the National Security Academy run by the military.

One important belief that is very much an integral part of the Turkish political culture, is the conviction that the external world is conspiring to weaken and carve up Turkey. Turkey is often depicted as surrounded by enemies who are very efficient and can act in unison. This phenomenon is often referred to as the 'Sèvres phobia', the fear that the

Treaty of Sèvres (drawn up by the victorious powers at the end of World War I and carving up the remaining Anatolian regions of the Ottoman Empire into small states and occupation zones) will be revived. The military plays a critical role in perpetuating the 'Sèvres phobia'. In his analysis of Middle Eastern politics Kocaoglu, a colonel in the Turkish army, provides an excellent example of how international affairs becomes one grand conspiracy when viewed through the lenses of such a phobia.[28] He tries to demonstrate how western policies in the Middle East are based on 'divide and rule' to ensure control over oil supplies. His analysis leads him to note that one primary goal of the United States in this conspiracy is to weaken and dismember Turkey. This would be an analysis with which many in Turkey would agree, without considering the fact that the United States has been an important military ally of Turkey since the US warship Missouri sailed into Turkish waters in 1946 as an expression of solidarity with Turkey against Soviet territorial demands. Hence, it is not surprising that the retired former Turkish Chief of General Staff, General Dogan Güres, does not think much differently either. In a recent interview he noted that he continues to suspect that it is the USA's intention to help the establishment of a Kurdish state in northern Iraq.[29] The general belief in Turkey is that such a state would threaten Turkey's territorial integrity.

The 'Sèvres phobia' permeates all walks of life. In 1995, the Turkish President, Demirel interpreted remarks made by the then French Minister of Foreign Affairs, Juppé during his visit that Turkey should find a political solution to the Kurdish problem as unequivocal evidence of Western intentions to create a Kurdish state in Turkey.[30] The French government's public announcement[31] that Juppé had never said or alluded to anything amounting to what Demirel was saying, did not weaken the grip of the 'Sèvres phobia' on Turkish public opinion. Many politicians and commentators filter developments in international affairs through the 'Sèvres phobia' and hence help to reinforce this belief among the public. A case in point is Ecevit, the leader of the influential Democratic Left Party (DSP). His belief in the 'Sèvres phobia' has pushed him on many occasions not only to oppose the continuation of Operation Provide Comfort but also to advocate the idea of restoring Iraq back to the international community under Saddam Hussein's leadership. The 'Sèvres phobia' is a critical factor that needs to be included in any effort to understand Turkish foreign policy towards countries with which Turkey may have disputes.

During the Cold War, by and large, the Ministry of Foreign Affairs and the military were the critical contributors to the government's foreign policy-making. It is possible to argue that during this period the government's pro-Western and status quo-oriented foreign policy commanded general support from political parties. The exception to this rule were the views of the Turkish Workers Party, which was critical of Turkey's close ties with the West and the United States in particular. Islamic and nationalist groups, which were then relatively weak compared to today, were pro-NATO which they considered to be anti-communist. However, in opposition to the general consensus they also advocated the idea that Turkey should play at the same time a leadership role in the Muslim world.

The end of the Cold War has created a new domestic political environment on foreign policy issues. The consensus on foreign policy issues has been weakened and a greater diversity of ideas are being expressed. The two conservative parties, Motherland (ANAP) and True Path (DYP), together with the social democratic Republican People's Party (CHP) advocate and support pro-Western policies and the continuation of economic integration with western Europe. Another social democrat political party, the Democratic Left Party (DSP), takes a more critical view of Turkey's relations with Europe and the United States. After the December 1995 elections, these three political parties commanded 392 out of a total of 550 seats at the Turkish parliament.[32]

It is the Islamic political party, the Welfare Party (Refah), which fundamentally opposed traditional Turkish foreign policy. Refah strongly objected to the ratification of the Customs Union between the European Union and Turkey. Instead, Erbakan, the leader of Refah, argued that Turkey should lead the formation of an Islamic United Nations, an Islamic Customs Union as well as an Islamic NATO.[33] Refah was very critical of Turkish foreign policy for being too passive on Bosnia–Herzegovina, the Azeri–Armenian conflict and the conflict in Chechnia. During the Imea/Kardak crisis he bitterly criticised Çiller for defining the problem as a problem of contested sovereignty and instead argued that the island belonged to Turkey. His forceful criticism played a role in influencing her to change stance and refer to the islets as part of Turkey rather than as islets with a disputed status. On relations with Iran, Refah has taken a position diametrically opposed to the one shared by other political parties and a large proportion of public opinion. Refah, by and large, refuses to lend support to frequent reports and criticisms of alleged Iranian support for terrorism in Turkey.

In April 1996, when a member of the parliament criticised the government for accepting a visit by the Iranian Foreign Minister soon after a crisis had erupted between Iran and Turkey over the confessions of an assassin of a Turkish journalist, implicating the involvement of Iranian diplomats, loud protests were heard from the ranks of Refah.[34] Refah members of the parliament used the occasion to accuse the government of being directly implicated in the suffering of civilians during the April 1996 Israeli operation against Hezbollah in southern Lebanon. The operation took place soon after a controversial military training agreement was signed between Israel and Turkey in February 1996, vehemently opposed by Refah. Refah consistently opposed the development and expansion of relations between the two countries.

In June 1996 Refah became the senior partner in a coalition government with the True Path party. Erbakan as the new Prime Minister very quickly recuperated from the setbacks caused by the extension of Operation Provide Comfort's mandate and the continuing warm relations with Israel by setting out on a long tour of Islamic countries in Asia, starting with Iran. The decision to choose Tehran as his first destination, coupled with the signing of a natural gas deal just at a time when additional US economic sanctions against Iran went into force, led to speculations that Turkey's foreign policy was changing in a fundamental manner. This was reinforced when Erbakan chose another group of Islamic countries in Africa, including Libya for a second major foreign policy tour within less than three months of coming to power. Erbakan did not travel to the West and in December 1996 he turned down an invitation to a dinner to be given at the end of the European Council meeting in Dublin.[35] Instead, early in January 1997 he hosted in Istanbul a meeting of ministers and officials from eight Muslim countries with populations over 60 millions that he hoped would lead to the formation of the group of D-8 (Developing-8).

The political parties that made up that governing coalition had almost diametrically opposed foreign policy priorities. However, the two parties appeared to have worked out a *modus vivendi*, as the Deputy Prime Minister and Minister of Foreign Affairs, Çiller, together with the President of Turkey, Demirel, continued to follow a more traditional Turkish foreign policy approach, emphasising Turkey's commitment to its relations with the West and its aspirations to integrate in Europe. Domestic political as well as ideological considerations pushed Erbakan to continue efforts to develop, in particular, economic relations with Islamic countries. However, there was not a major shift in Turkish foreign policy. Structural, institutional and domestic political factors stood in the way of such a drastic change. Instead, as long as Refah remained in power Turkey's foreign policy agenda and relations became more multi-dimensional and diverse.

One last factor that should be raised in terms of foreign policy-making in Turkey is that the role of interest groups has been increasing. This became particularly visible during 1995 in respect to the debate concerning the Customs Union with the EU. There were many important business associations as well as pro-European groups who actively lobbied the government as well as political actors across Europe, to mobilise support in favour of the Customs Union. Business interest groups have also become an important factor in Turkish–Russian relations. Business associations and various civic organisations are also becoming increasingly vocal in stressing the need to develop better relations with Greece. Similarly, ethnically and religiously based groups have also become active on many foreign policy issues. For example, the government often found itself trapped between the demands of such groups in respect to the Chechen conflict and business associations representing large commercial interests in Russia. Lastly, both domestic as well as international human rights organisations increasingly influence Turkish foreign policy-making, particularly on issues related to the Kurdish problem in Turkey.

## THE EUROPEAN UNION LEVEL

An important factor that guides Turkish foreign policy towards the European Union is intricately related to the legacy of the founders of modern Turkey. For them the transition from the theocratic Ottoman Empire to a modern and secular Turkish Republic was to be achieved through westernisation. Modernisation was very much understood to be a process of emulating and eventually becoming part of western civilisation as a European state.[36] The emergence of the Cold War and the Soviet threat added an important security dimension to this process and heightened the urgency to become part of the West. At the same time Turkey's strategic importance for the West facilitated Turkey's entrance into institutions like the Organisation for Economic Cooperation and Development, the Council of Europe and NATO.

The next major step came when the Turkish government applied in 1959 for an association agreement with the European Economic Community (EEC). The military intervention in 1960 led to a delay in the negotiations but by 1963 the Ankara Treaty was signed. The agreement envisaged the gradual establishment of a customs union between Turkey and the EEC that would eventually evolve into a full economic union. Turkey's motivation behind signing such an agreement was far from being purely economic. As Vali noted, it was also very much driven by political considerations and the desire not to be left behind Greece which had already signed its own association agreement with the EEC.[37] However, the factor that needs to be emphasised is that among the Turkish elite this agreement was considered to be part of the process started by their predecessors to join the Western world. As Eralp notes, 'According to Turkish policy-makers, membership of the new European Community (EC) was a logical extension of Turkey's inclusion in the other Western organisations'.[38]

In economic terms the agreement did play an important role in that the EC has become Turkey's most important trading partner. However, a number of problems complicated the implementation of the Ankara Treaty. The oil crisis led the Turkish economy to face serious problems, particularly in the second half of the 1970s, which affected Turkey's implementation of its obligations under the Treaty. The situation was exacerbated by increasing political instability and violence between extremist groups within Turkey. Economic difficulties struck members of the EC too. Germany began to demand revisions to the provisions of the Treaty concerning the free movement of Turkish labour, while France and Italy sought revisions concerning Turkish agricultural exports. Turkey's relations with the EC entered a crisis period when the military intervened in Turkish politics once more in 1980. The relations remained more or less frozen until the mid-1980s when Turkish democracy began to recover.

The 1980s can be characterised as a period during which the economy went through major structural changes. The inward-looking Turkish industry based on import-substitution policies was replaced by an export-driven one. This was accompanied by an extensive liberalisation of the Turkish economy. Between 1981 and 1987, the Turkish economy expanded by an average growth rate of 5.5 per cent.[39] Turkish foreign trade also grew in a very significant way. Even though during this period the share of the EC's foreign trade fell as Turkey's trade with the Middle East expanded, with the exception of the period between 1981–1983 the EC still remained Turkey's largest trading partner.[40] The growth in the industrial sector of the economy encouraged the industrialists to support the idea of full membership.[41] The period also saw greater democratisation taking place in Turkey. A growing confidence in Turkey's economic performance and its democracy accompanied with the personal ambitions of Özal, the then Prime Minister, to follow a more outward looking foreign policy culminated in the Turkish application for membership to the EC in 1987.

By and large the application received considerable public support in Turkey. A 1987 public opinion poll indicated more than 61 per cent support for the government's efforts to join the EC. Similarly, with the exception of right wing religious and nationalist political parties, other parties were supportive of the idea,[42] which was also supported by the foreign policy and military elites who saw it as a natural extension of Turkey's westernisation efforts. The European Commission rendered its decision in December 1989 and concluded that there were major fundamental economic and political problems that needed to be resolved before Turkey's admission could become possible. The decision led to considerable disappointment in Turkey and many attributed the decision to a European reluctance to accept Turks into their ranks.[43]

The Commission's decision coincided with the Berlin Wall coming down and the Cold War beginning to wane. This led to growing concerns that Turkey, besides being turned down by Europe, was also loosing its strategic importance for the West. The collapse of the Soviet Union and the emergence of independent states around the Black Sea and Central Asia somewhat made up for the disillusionment. Özal, as the new President of Turkey, was quick to exploit these developments to compensate for the loss of morale by introducing slogans such as 'the next century, the twenty-first century, would be the century of Turks' and 'a Turkish world unfolding from the Adriatic to the Chinese Wall'. These slogans helped to soften the impact of what many considered to be the exclusion from Europe and led many to consider replacing Europe with closer economic, cultural and political links with Turkic countries.[44] This was accompanied by the idea of developing stronger economic ties between countries around the Black Sea to tap a market of 350 million consumers. This opened the way to the establishment of the Black Sea Economic Cooperation in 1992.

The 1990 Iraqi invasion of Kuwait and the Turkish decision to support the operations against Iraq significantly altered the Turkish foreign policy-making environment. As Sayari rightly notes, the Gulf crisis 'erupted at a time when Turkish perceptions about the restructuring of the European security system were tinged with considerable pessimism and concern'.[45] In the words of Brown, the critical role that Turkey played during the Gulf crisis 'revitalised Turkey's strategic importance, bordering as it does the Middle East'.[46] In this new environment Turkey's secular regime and its relative economic success began to be portrayed as an important source of stability. Many also went as far as to suggest that Turkey could become a model for the newly independent Muslim countries emerging from the dissolution of the Soviet Union.[47] Furthermore, European interest both in the Turkish economy and in the possibility of joint ventures in Central

Asia increased. Similarly, the US administration included Turkey among the 'Big Emerging Markets'. This new post-Gulf crisis environment was critical in reinforcing the Turkish government's commitment to see Turkey's continued integration with Europe. It is against such a background that in March 1995 the Customs Union agreement between Turkey and the EU was signed.

However, the agreement required the ratification of the European Parliament. In the course of the past few years the European Parliament had become highly critical of the Turkish government's human rights violation, especially in respect to the Kurdish problem. The forced evacuation of villages in southeastern Turkey and the military's heavy-handed approach to the struggle against the PKK did not help. Matters became even worse when the parliamentary immunities of thirteen deputies from the Kurdish Democracy Party (DEP) were lifted in March 1994 and seven of them were convicted to various prison terms in December. Another difficulty that the customs union faced was Greek objections to the development of closer relations between Turkey and the EU.

Against this background during the course of 1995 the Turkish government came under intense pressure from EU member governments to introduce political reforms. However, at the same time the governments of leading members of the EU, such as Britain, Germany, France, Italy and Spain, also demonstrated their support for the signing of the Customs Union with Turkey. Expanded economic relations with Turkey was considered as an important way of ensuring the stability of the Turkish regime. There was a recognition that a stable democratic and secular Turkey would be more likely to contribute to European security than a Turkey that might turn increasingly eastward. The expanding Turkish economy and a market of 60 million consumers offered prospects of greater trade and business. The government of the United States and Israel lent strong support for the ratification of the Customs Union agreement.

Domestic political development in Turkey also helped the emergence of an environment conducive to the ratification of the Customs Union. In July 1995 the government succeeded in getting through parliament the first ever package of amendments to the 1980 military era Constitution.[48] Most significantly the amendments made it more difficult to lift the immunities of members of parliament and opened the way to a freer civil society by facilitating the formation of, and membership to, civil associations. Another important breakthrough in terms of the Kurdish question came late in October 1995 when the Turkish Parliament accepted new legislation amending the infamous Article 8 of the Anti-Terror Law.[49] The amendments provided for a reduction in prison terms for offenders and a tighter definition of 'separatist' propaganda. The latter was particularly important because previously anybody calling for a political solution to the Kurdish issue ran the risk of punishment. The amendments were also made to apply retroactively allowing the courts to free some of the previous offenders. Though many critics found the changes somewhat cosmetic, they opened the way to a relatively freer discussion of the Kurdish question in Turkey.

International support and improvements in the Turkish legal system played a critical role in the eventual decision of the European Parliament to ratify the Customs Union agreement in December 1995. However, since the Customs Union came into effect on 1 January 1996 the Turkish government has been critical of the EU for failing to live up to its obligations. In March 1996 one of the bodies of the Customs Union, the Association Council, which is vital to the overseeing of the implementation of the agreement, failed to meet in the face of Greek objections. The Fourth Financial Protocol that was initialled back in 1980 and earmarked credits amounting to 600 million ECU had not come into force due to similar objections. In October 1996 another disillusionment for Turkey was the European Parliament's decision to freeze EU assistance to Turkey on the grounds of a failure by the Turkish government to meet its human rights promises. The decision attracted fierce criticism both from officials as well as business interest groups and civic organisations in Turkey.[50]

Eralp's remark that Greece has been able to use the Community platform to voice her demands and cast her veto on EC–Turkish cooperation continues to hold.[51] Therefore, the many bilateral problems between Greece and Turkey are likely to continue to obstruct the smooth development of the customs union. The question of Cyprus's eventual accession to the EU is also likely to complicate relations. The many problems between the EU and Turkey also erode Turkey's commitment and support for integration in Europe. Lesser has argued that this might lead the elite in Turkey to look for opportunities outside Europe.[52] Of course, while the government led by the Islamic Refah was in place this trend might have been reinforced. However, in general there are also clear signs that Turkey remains committed to the customs union and integration with Europe.

This was reiterated by the then Turkish Foreign Minister during his address to the Turkish Parliament in April 1996. The snub that Turkey received from the EU when it failed to include Turkey among the potential future member countries invited to the Turin meeting of the Intergovernmental Conference in March 1996 did not change the government's determination. This was reflected in the statement made by the then Turkish Prime Minister, Yilmaz, that 'this would be the last meeting without Turkey'.[53] The coalition government led by Erbakan remained committed to the implementation of the customs union even though he personally had opposed the ratification of the treaty and had promised that if Refah was to come to power he would at least try to renegotiate it. As Hale notes, it is significant that that coalition government's programme confirmed Turkey's commitment to fulfil the targets of the Customs Union agreement. Though Hale does add that this change in Erbakan's position might well be explained by the fact that Refah was well short of an overall parliamentary majority.[54]

The President of Turkey, Demirel, too remains committed to the Customs Union and to Turkey's integration with the EU. During a visit to Italy in October 1996 he reiterated Turkey's aspiration to become a member of the EU and sought the assistance of Italy to achieve this goal. The significance of this visit was enhanced by the fact that it coincided with Erbakan's controversial visit to Libya. In January 1997, Çiller met with the Foreign Ministers of the five leading members of the EU in Rome. She reiterated her commitment to see Turkey as an eventual member of the EU and sought their support for this end. Also in a dramatic manner, she raised the stakes in European–Turkish relations when she linked the issue of NATO's expansion to closer ties between the EU and Turkey by raising the question 'How do you think you can expand NATO without Turkey's permission?'.[55]

Turkish public opinion also seems to be supportive of the government's aspirations of integration with Europe. This was reflected in an opinion poll in December 1995 which showed that 72.6 per cent of those polled said yes to the question 'Do you support Turkey's entry into the customs union?'.[56] However, overall Kramer is right in concluding that future relations between Turkey and the EU are set to become more diverse and complex and that the real test for these relations will become much sharper now that the customs union has come into effect.[57] This assessment was vindicated by the policies of the new coalition goverment with Yilmaz as Prime Minister and Cem as Foreign Minister, which replaced that of Erbakan in June 1997.

## ECONOMIC, FINANCIAL AND TRADE ISSUES

The importance that Turkey attributes to its relations with the EU can also be explained in terms of trade and economics. As Table 9.1 shows, the EU continues to be Turkey's most important trading partner. Furthermore, the trading relationship is very complex and multidimensional. Although in the past Turkish exports were basically agricultural products and textiles, more recently Turkish exports to the EU have become more diversified. At the same time, Turkey with a growing middle-class and urban population is becoming an attractive market for the EU. Economic liberalisation policies in the 1980s have significantly transformed the Turkish economy from a closed economy to a relatively open one. The Customs Union can partly be seen as a manifestation of this development. However, what is striking is that it is the non-Mediterranean countries of the EU that have close economic relations with Turkey: Britain, Germany and The Netherlands are more important economic partners than France, Greece, Italy and Spain.

A similar trend can also be observed in respect to foreign direct investments. As Table 9.2 shows, during the course of 1995 there were 1,225 investment permits issued by the

*Table* 9.1 Turkish foreign trade by geographical regions for Jan–Nov 1994 and Jan–Nov 1995 in billion $[58]

| | *1994* | | | | *1995* | | | |
|---|---|---|---|---|---|---|---|---|
| | *Exp.* | | *Imp.* | | *Exp.* | | *Imp.* | |
| | | % | | % | | % | | % |
| EU (Med.) | 2.0 | (13.6) | 3.5 | (17.0) | 2.8 | (14.4) | 5.4 | (17.0) |
| EU (Non-Med.) | 5.6 | (35.4) | 6.1 | (29.9) | 7.2 | (37.3) | 9.4 | (30.1) |
| Med. (Non-EU) | 0.8 | (5.1) | 0.6 | (2.7) | 1.0 | (5.3) | 1.2 | (3.9) |
| Black Sea | 1.1 | (7.1) | 1.7 | (8.4) | 1.8 | (9.5) | 3.3 | (10.6) |
| Central Asia | 0.2 | (1.5) | 0.2 | (0.8) | 0.3 | (1.5) | 0.2 | (0.8) |
| USA | 1.3 | (8.2) | 2.2 | (10.7) | 1.4 | (7.0) | 3.1 | (10.0) |
| Japan | 0.1 | (0.9) | 0.8 | (3.9) | 0.1 | (0.7) | 1.2 | (3.9) |
| Others | 4.5 | (28.7) | 5.5 | (26.6) | 4.7 | (24.2) | 7.5 | (23.8) |
| Total | 15.8 | | 20.0 | | 19.2 | | 31.4 | |

*Sources*: *Baslica Ekonomik Göstergeler* (Basbakanlik Dis Ticaret Müstesarligi, Ankara, Subat 1996), p. 60; *Aylik Ekonomik Veriler* (ITO, Istanbul, Nisan 1996), p. 35; *Temel Ekonomik Göstergeler* (DPT, Ankara, Subat 1996), p. 68.

*Table* 9.2 Foreign direct investment in Turkey by geographical regions and countries for 1995[62]

| | *No. of permits* | *Amount of capital in million $* | % |
|---|---|---|---|
| EU (Med.) | 142 | 580.6 | 19.8 |
| EU (Non-Med.) | 424 | 1256.1 | 42.7 |
| Med. (Non-EU) | 24 | 16.0 | 0.5 |
| Black Sea | – | 0.5 | 0.02 |
| USA | 71 | 231.4 | 7.9 |
| Japan | 19 | 283.8 | 9.7 |
| Others | 545 | 569.9 | 19.4 |
| Total | 1225 | 2938.3 | |

*Sources*: *Baslica Ekonomik Göstergeler* (Basbakanlik Dis Ticaret Müstesarligi, Ankara, Subat 1996), p. 44; *Aylik Ekonomik Veriler* (ITO, Istanbul, Nisan 1996), p. 15.

Turkish government for foreign capital. This amounted to a total of just under $3 billion. More than 50 per cent of this capital came from EU members, with the largest proportion coming from EU countries outside the Mediterranean basin, although France, with $476 million, was the second largest foreign investor in Turkey. The total investments of Portugal, Spain, Italy and Greece in 1995 barely exceeded $100 million. Among the EU countries in 1995 The Netherlands was at the top of the list of foreign investors in Turkey with $560 million. The 'Others' group includes Switzerland, Japan and the United States as the major investors while Middle Eastern and Islamic countries' investments amounted to less than $70 million.

Both trade and foreign investment statistics show that the Turkish economy remains closely tied to the EU. The Customs Union will reinforce this trend although it should be noted that Turkey's trade with Black Sea countries and particularly with Russia is steadily increasing. There is also a booming 'suitcase' trade taking place between Turkey and Russia as well as other CIS countries totalling many billion dollars.[59] Furthermore, Russia has also become a very important market for Turkish construction companies. Turkish companies are actively involved in construction projects in Russia amounting to $2.7 billion and they have completed work valued at another $2.2 billion.[60] Ongoing contracts in Russia constitute almost half of the total work in progress worth approximately $6.2 billion. Clearly, Russia together with other CIS countries have replaced the Middle East as the most important market for Turkish construction companies.[61]

Turkish economic relations with Arab countries increased significantly during the 1970s and 1980s.[63] This changed with the Gulf crisis in 1990 and the UN trade embargo on Iraq. This embargo has also affected transit trade. As a result Turkey not only lost access to the Iraqi market but also to Arab countries beyond Iraq. By and large, Arab–Turkish trade has been reduced to Turkish purchases of oil. In 1994 Turkey purchased 21.2 million metric tons of oil, most of it from Iran and Arab countries.[64] However, even during the peak of Arab–Turkish economic relations Arab countries along the Mediterranean did not have close economic ties with Turkey, with the exception of Libya. The situation is not very different today. As is shown in Table 9.1, only a very small proportion of Turkish foreign trade is with non-European Mediterranean countries. Furthermore, economic relations with Libya have also been adversely affected by UN sanctions on Libya.

In contrast to Arab countries trade with Israel is expanding. The Arab–Israeli peace process has opened the way for Turkey to develop its relations with Israel, which had been passive since the 1950s. In 1995 Turkish trade with Israel amounted to almost half a billion dollars. This trade was expanding fast and was expected to grow even faster as the free trade agreement signed between the two countries in March 1996 took effect. Israeli tourism in Turkey is growing and Israeli companies are taking growing interest in the Southeastern Anatolian Development Project for investment opportunities. Turkey is also aiming to expand its economic relations with the Palestinians. Turkey has pledged 50 million dollars of credit for the Palestinian Authority and is very much interested in bidding for infrastructural projects. However, so far Turkey has failed to release this credit and not much has happened in terms of Turkish–Palestinian economic relations.

In 1994 the Turkish government attended the Casablanca summit on regional economic cooperation with great enthusiasm. However, this enthusiasm was not sustained for long. Instead, at the October 1995 Amman summit the Turkish delegation did not even include any members of the government, let alone the prime minister, as had been the case at Casablanca. The enthusiasm for a Mediterranean role was short-lived although this situation somewhat improved at the Cairo summit in November 1996, where the Turkish government was represented at the ministerial level and supported by a large group of business people. Turkey also participated at the Euro-Mediterranean Conference in Barcelona in November 1995. From the point of view of the Turkish decision-makers the Conference was important in respect of the EU's commitment to make available financial assistance amounting to 4.7 billion ECU between 1995 and 1999. The decision to establish a free-trade area around the Mediterranean by the year 2010 was a decision that remained remote to Turkish interests, although in the long run the idea of a free-trade area might become attractive to Turkey in terms of developing closer trade and economic relations with Tunisia and Morocco.

## FOREIGN POLICY

In the new post-Cold War environment ethnic conflicts have become one of the most difficult challenges that Turkish foreign policy-makers have had to face. Ethnic conflicts in the Balkans and the Caucasus have led to Turkish involvement, partly due to domestic pressures, partly to a perceived need to contain these conflicts from escalating in a manner that could draw Turkey into them. In the Balkans Turkey called for an interventionist policy against the Serbs in Bosnia-Herzegovina and bitterly criticised Western governments for failing to stop the aggression and prevent the dismemberment of the country. However, the government refrained from taking unilateral action, which was being demanded by radical Islamic groups in Turkey. Instead, Turkey contributed air and naval units to NATO forces as well as ground forces to UNPROFOR. At the same time the Turkish government was active in encouraging the formation of a military alliance and then a federation between Croats and Muslims. The signing of the Dayton peace agreements in November 1995 fell short of Turkish expectations but was nevertheless received with a relief. The end of the conflict opened the possibility for Turkey to rebuild its relations with Serbia,[65] a country that has a critical location in terms of Turkish land trade and movement of people towards the EU.

In respect to ethnic conflicts in the Caucasus, the Turkish government is often trapped between two forces. One force is composed of radical Islamic groups and ethnic activist groups of Caucasian descent that have traditionally supported secessionist movements in the land of their ancestors. The other force is a product of a combination of factors that compel the government to follow a more balanced approach to regional ethnic conflicts. First, the Kurdish problem in Turkey and the ethnic conflicts in the Caucasus can be paralleled. The Turkish government's concern about external involvement in the Kurdish problem and the activities of the separatist PKK obliges it to refrain from overtly supporting separatist movements in the Caucasus. Second, large Turkish companies and business associations with economic interests in Russia have pressurised the government to maintain good relations with Russia. For example, the Director of Alarko Holding, Alaton, noted that their business dealings in Russia were being adversely affected because of the Chechen conflict and that the Turkish government should be more sensitive to Russian concerns.[66] Akgün too has noted that increasing economic relations and growing economic interdependence between Turkey and Russia, call for the Turkish government to develop better relations with Russia.[67]

In the case of Abkhazians in Georgia, in spite of domestic public opinion pressures the Turkish government has been supportive of the Georgian government and has maintained what the President of Georgia during his last visit to Turkey called 'excellent relations'.[68] The common fear of growing Russian influence in the area as well as common interests in respect to oil-pipeline issues are an important factor that draw the two governments closer. The case of the Chechen conflict is less clear cut. The groups supportive of the Chechens in Turkey are much more vocal. The death of the rebel Chechen leader Dudayev in April 1996 led to anti-Russian demonstrations. As Winrow notes, the linkages between the Chechen and Kurdish problem make the Turkish government very wary of open support for the Chechens.[69] The successful negotiations by Yeltsin's former advisor on security affairs, Lebed, in early September 1996 and the presidential elections held in Chechnya in January 1997 have to an important extent eased the domestic political pressures on Turkish foreign policy-makers to support Chechen rebels.

It is the conflict between Armenians and Azeris that constitutes the greatest dilemma for the Turkish government. The ethnic and cultural links between Azerbaijan and Turkey weigh heavily on policy-makers. This makes it very difficult for the Turkish government to improve its relations with Armenia when one-fifth of Azerbaijan's territories are under Armenian occupation. Yet, at the same time there are also powerful forces that push the government towards searching for ways of coming out of the impasse. The Turkish provinces bordering Armenia have a very depressed economy and there are local pressures to open the border between

Armenia and Turkey to encourage local trade. More importantly, the powerful Armenian lobby in the US Congress is very active in efforts to try to introduce legislation against Turkish interests. This threatens to disturb Turkey's relations with the US government, a relationship that both sides seem to value highly. In an effort to win the goodwill of the US government the then Turkish Prime Minister, Çiller, announced during her visit to the US in April 1995 the decision to reopen air-corridors for humanitarian assistance purposes.[70] Another gesture towards Armenia came when the Turkish government expressed its readiness in February 1996 to open the land border if there were an Armenian public and unequivocal commitment of an eventual withdrawal from Azeri territories in return for autonomy for the Armenians in Nagorna-Karabagh. Previously, the government used to insist on complete and effective withdrawal before relations could be started.

The dissolution of the Soviet Union had come as a great relief and joy to Turkey. The relief was a product of a Russian military withdrawal from the borders of Turkey and resultant diminishing of an immediate security threat. The joy came when five independent Turkic states emerged from the ex-Soviet Union. This led to considerable euphoria in Turkey and expectation of close economic and political relations. However, this sense of relief and joy did not last very long. Russian efforts to reintroduce its influence in the 'near abroad' brought the Russian military back to the Turkish border with Armenia and Georgia. The Russian decision not to meet the deadline for the reduction of forces in the Caucasus set in the 1990 Conventional Forces in Europe Treaty led to Turkish objections and fuelled Turkey's sense of insecurity. In May 1996 during his visit to Uzbekistan the Turkish President, Demirel, criticised the agreement for economic and cultural integration signed between Russia, Belarus, Kyrgyzstan and Kazakhstan. He expressed his fear that this might be a step in an effort to revive the Soviet Union.[71]

Turkey's relations with Central Asia has equally led to some disappointments. The euphoria and the expectations of developing close relations with Central Asian countries met serious pragmatic and political difficulties. Although Turkey's economic relations with these countries have expanded significantly, Turkey's limited resources have meant that Turkey has not been able to meet the developmental aspirations of these countries. Central Asian countries have had to continue to rely on their close relations with Russia but also try to expand relations with China, Japan and Iran. This has led Robins to argue that the leadership of Central Asian countries have had to prefer self-interest in place of sentiment.[72] Similarly, in terms of closer political relations Özal's dream of achieving a Turkic Union did not survive beyond the first summit of Turkic states in 1992. The ambitious draft declaration prepared by Özal had to be significantly watered down because, as Winrow notes, 'Central Asians were eager to cultivate closer political and economic relations with other states in addition to Turkey' and furthermore 'were also careful not to provoke Moscow by binding themselves too closely to Turkey'.[73] The latter is a particularly potent concern considering the then Russian Foreign Minister Kozirev's remark 'we do not want a political union established on the basis of the Turkish language'.[74]

In recent years the question of the routing of Caspian Sea and Central Asia oil has been an important item on the Turkish foreign policy agenda. A number of different routes are being discussed. The Turkish government preference has been for a route that would cross Turkey from the Caucasus to the Mediterranean coast. The reasons have been a mixture of environmental, economic and political concerns. Turkey fears that if oil is transported to the Russian port of Novorossiysk and then shipped in super tankers through the Turkish Straits this would increase the likelihood of accidents with dire environmental consequences. The construction of an oil pipeline going through Turkey would make a significant economic contribution to the economy of a region of Turkey that is underdeveloped. However, most importantly, Turkey hopes that if such an oil pipeline were to be built this would prevent Russia from enjoying political influence over oil-exporting Central Asian countries. The decision reached in April 1996 to carry Kazakh oil through Russia has come as a major setback to Turkish plans.

In the Middle East, as mentioned above, Turkish foreign policy has become much more varied and dynamic than it was in the Cold War years. However, relations between Turkey and Iraq have remained frozen since Iraq's invasion of Kuwait and the Turkish decision to support the military operation of the Allied forces. This has been complicated by the role that Turkey has played in the creation of a Kurdish safe haven in northern Iraq in April 1991. Since then by allowing Operation Provide Comfort to continue, Turkey has prevented Iraq from reasserting its sovereignty in the area. In the meantime the Kurdish groups have declared a federated Kurdish state and have been trying to consolidate their power there. The ability of the PKK to operate from these Kurdish- controlled parts of northern Iraq has led the Turkish military to mount numerous operations into the area.[75] These operations as well as Turkey's involvement in the conflict between Kurdish groups in northern Iraq has drawn regular criticism from the Iraqi government.

Turkey has also played a critical role in maintaining the trade embargo against Iraq at some cost to itself. As a result of the embargo Turkey has lost its trade with Iraq as well as with Arab countries beyond Iraq. Turkey also lost the revenues from the Kerkuk–Yumurtalik oil pipeline that used to carry Iraqi oil to the Mediterranean Sea. These losses are put at approximately $25 billion. It was the desire to recover some of these losses that led the Turkish government in 1994 to try and persuade Iraq and the Security Council members to accept a partial lifting of the embargo by allowing Iraq to sell a controlled amount of oil through Turkey in return for the purchase of food and medicine. The Security Council in April 1995 adopted Resolution 986, incorporating some of these ideas. After very long and arduous negotiations in May 1996 Iraq finally agreed to cooperate with the United Nations for the implementation of the terms of this Resolution. The Iraqi government's military involvement in late August 1996 in the conflict between Kurdish groups in northern Iraq caused delay in the implementation of this Resolution until December 1996. According to this Resolution, every six months Iraq would be permitted to export $2 billion of oil some of which would be transported through Turkey. Turkish business was also expected to benefit from Iraqi imports of food and medicine.

It is Turkey's relations with Greece and the problem of Cyprus that continue to constitute one of the greatest challenges to Turkish foreign policy. The problems between Greece and Turkey are numerous. They range from long-standing conflicting claims over the status of the Aegean Sea to the treatment of their respective Greek and Turkish minorities. More recently, Greek policies have also affected relations between the European Union and Turkey. It is possible to say that the two countries suffer from a deep lack of trust. The problems between the two countries are also complicated by domestic political considerations.

However, in a significant break from past practice soon after he took office, the then Turkish Prime Minister, Yilmaz, announced that the Turkish government would be prepared to accept third-party arbitration to solve problems of a legal nature between the two countries. Previously, while Greece argued for the need to go to the International Court of Justice, Turkey preferred political dialogue and negotiations. Rightly, Kut considers this as a major and significant psychological shift in the Turkish position.[76] There are also some positive developments in Greece that might lead to improved relations. In September 1996 the moderate Greek leader, Simitis, formed a new government after PASOK won early national elections. The new government was welcomed in Turkey and the Turkish Prime Minister immediately called for the development of friendly relations through a process of constructive dialogue.[77] A similar call was also made by the Turkish President in October during his visit to Italy.

According to Bagci, the fact that Greek political parties which based their campaigns on anti-Turkish slogans did not perform well at the elections and that Simitis did not include the former hawkish Defence Minister Arsenis in his cabinet were positive signs for better relations between the two countries.[78] Akgün too noted that statements made by the Greek Foreign Minister, Pangalos, suggesting a dialogue with Turkey were another positive development but that unfortunately these had gone unnoticed in Turkey.[79] However, as the participants at a seminar on Greek–Turkish relations in Istanbul in December 1996 recognised, the crucial factor still remained the question of which government should make the first concession to start a dialogue.[80] It is difficult to see how a weak coalition government in Turkey might be able to muster the necessary political will for concessions.

Relations between Greece and Turkey are also a hostage to the Cyprus problem. The violence between the two Greek and Turkish communities and the coup d'état perpetrated by an officer of the Cypriot National Guard, Sampson, against President Makarios led to a Turkish intervention on the island in 1974. Since the 1974 intervention the two communities have been cut apart and the Turkish side unilaterally declared independence in 1983. The efforts of the United Nations to find a solution acceptable to both communities have failed. Hence, it is also difficult to see how Greek–Turkish relations could improve without progress towards the solution of the Cyprus problem. This difficulty was demonstrated in a dramatic manner when the Turkish government in January 1997 threatened to use force if Russian S-300 ballistic missiles purchased by Cyprus were actually deployed. The news about the missiles and the Turkish reaction increased the tension between Turkey and Greece.

## SECURITY CHALLENGES

During the Cold War, Turkey's security was connected closely to that of the West. This was ensured through Turkey's membership of NATO and Turkey's strategic importance to the defence of the West. However, this relatively neat security arrangement has changed significantly since the end of the Cold War. Mortimer has identified two zones of unbroken instability 'the first stretching from the Arctic Circle down through Eastern Europe, the Balkans and the Middle East, and the second along the southern shore of the Mediterranean'.[81] It is not difficult to notice that Turkey straddles the heart of the area where the two zones overlap. A report prepared by the Western European Union (WEU) also recognised the unique position of Turkey at the centre of such an unstable region.[82]

It is possible to talk about two types of security problem for Turkey. The first can be considered to be the more classic type, depending on a narrow definition of security: the protection of Turkey from risks that Turkish decision-makers perceive as threatening the very existence of Turkey as they know it. Such a definition is associated with the physical protection of central values such as independence and territorial integrity of the country and is closely associated with military security. The second type of security problem depends on a much broader concept, one that envisages much more ambiguous threats emanating from all kinds of sources, ranging from ethnic rivalries to mass migrations due to economic, political and social upheavals. Security from environmental disasters would also be included in such a definition. After all, as Baldwin points out, in order to survive 'states need minimum amounts not only of security from external attacks but also of breathable air, drinkable water, economic welfare, and so forth'.[83]

Turkey faces security challenges of both types. In respect to the first, it is possible to argue that Turkey continues to value highly its membership to NATO as a means of deterring external military threats. However, as in the rest of Europe, there is a recognition that NATO alone cannot adequately ensure security. The failure of the United Nations and the OSCE in stopping the violence in Bosnia–Herzegovina and the inability of the OSCE to resolve the Armenian–Azeri conflict have been very disappointing. Yet, at the same time the Turkish government has shied away from supporting any unilateral interventions. Instead, the government's policy has been based on the need to use international organisations to resolve, or at least to manage, disputes that threaten to spill over into large armed conflicts. At the same time Turkey has been supportive of the idea of interlocking European and political organisations expressed in the Rome Declaration of NATO of November 1991.[84]

However, since the Rome Declaration there has been a growing unease about the reliablity of Turkey's defence relations with the West. As Snyder notes, a combination of factors, including 'a reduced US presence in and commitment to Europe's security, lack of western resolve in addressing the Balkan crisis, and an increasingly visible and aggressive Russian military presence in the Caucasus region, suggest to Turkey that its iron-clad links to NATO and the west are more fragile than they have been in several decades'.[85] In Turkey this is also accompanied by a sense of becoming marginalised as a western ally at a time when the eastwards expansion of NATO and the EU is gathering momentum.

The discomfort felt by this situation was very much reflected in the declaration of Çiller, then Turkish Minister of Foreign Affairs, during a meeting of the Western European Union (WEU) in November 1996 that Turkey would veto the use of NATO forces in Combined Joint Task Forces run by the WEU. In February 1997 during the visit of NATO's Secretary General, Solana, Turkish officials also threatened to use their veto against the expansion of NATO's membership if closer ties were not developed between Turkey, the EU and the WEU.

Presently, the most potent threat to Turkish security comes from the armed activities of the PKK. The confrontation between the Turkish security forces and the PKK has led to the death of more than 20,000 people and the displacement of large numbers of people. As a part of its campaign to establish an independent breakaway Kurdish state, the PKK has mounted numerous acts of terrorism ranging from bomb attacks on tourists to armed attacks on vehicles carrying civilians . As a result an important proportion of Turkey's military effectives have been redeployed in southeastern Anatolia. As Özel points out this is casting a shadow on Turkey's ability to deter potential external attacks.[86]

The sophisticated military capabilities of neighbouring Middle Eastern countries and their past war-making record is considered by the government to be another important threat to Turkish national security. The fact that Iran and Syria are equipped with ballistic missiles and that there are reports about Iranian efforts to acquire a nuclear capability are serious sources of concern. As Karaosmanoglu notes 'some of Turkey's important population centres, dams, power stations, airbases, and military headquarters are within range of these systems'.[87] Although presently Iraq does not constitute an immediate military threat to Turkey, most Turkish decision-makers still vividly recall Saddam Hussein's belligerent posture against the visiting Turkish Prime Minister, Akbulut, only three months before his invasion of Kuwait. At the time Saddam Hussein had told him 'NATO is disintegrating. Your friend, the United States is losing power... Nobody listens to the US anymore. She cannot help you'.[88] Given Turkey's critical role in getting Iraq out of Kuwait and the subsequent events, and unless there is a major change in the

Iraqi regime, an Iraq armed with sophisticated weaponry will continue to constitute a major threat to Turkey. The potential military threats from these countries were critical in Turkey getting the participants at the CFE negotiations to accept that the southeastern regions of Turkey be excluded from the coverage of the CFE Treaty.

Among the many problems between Turkey and Greece it is the question of the width of territorial waters in the Aegean that Turkish decision-makers consider to directly affect Turkish national security. Greece is a signatory of the UN Convention on the Law of the Sea. The Convention allows countries to extend their territorial waters to 12 miles. In Turkey there is great apprehension about Greece extending its territorial waters. Such an extension would increase Greek territorial waters from 43.8 to 71.5 per cent of the Aegean.[89] Most importantly this would reduce the size of the high seas in the Aegean Sea from 48.9 to 19.7 per cent leaving no high seas between the Turkish Straits and the Mediterranean Sea. After the Greek ratification of the Convention in June 1995, the perceived threat led the Turkish Parliament to declare any extension of the territorial waters as 'unacceptable and granted the government all powers deemed necessary to defend the country's interests'.[90] The decision in turn further soured already tense Greek–Turkish relations and played an important role in paving the way to the Imea/Kardak crisis in January 1996.

The new challenges to Turkish national security since the end of the Cold War have put Turkey in the most unusual position of being one of the few countries continuing to maintain high national defence budgets. After Greece, Britain and the United States, Turkey had the largest defence budget in proportion to its GDP in 1994.[91] Turkey started to manufacture F-16s and in 1995 it leased US tanker planes and is negotiating the purchase of AWACS to develop a deterrence capability, particularly against ballistic missiles in the Middle East. In the post-Cold War era Turkey also became the first NATO member country to purchase ex- Soviet military equipment from Russia. In the face of growing US Congressional opposition against the sale of sophisticated helicopters Turkey turned to Russia and France to buy helicopters to combat the PKK in mountainous terrain. It has been reported that Turkish armed forces will need resources amounting to $150 billion for its weapon requirements in the coming 25 years.[92]

Turkey has also been directly involved in modest efforts to encourage cooperation on security among Middle Eastern countries since the Arab–Israeli Peace Process started in October 1991. The most significant one was probably the decision to contribute Turkish troops to the international force enlarged in Hebron after the agreement reached between Israel and the PLO in January 1997. In May 1993 in the context of the efforts of the Multilateral Working Group on Arms Control and Regional Security to strengthen the larger Middle Eastern peace process, Turkey was given the task of guiding the workshop on 'Military Exchanges of Information and Prenotification of Certain Military Activities'. In this respect the Turkish Ministry of Foreign Affairs introduced a draft package of measures that were discussed in a series of meetings by the members of the workshop. In a plenary meeting in Tunisia in December 1994 the Multilateral Working Group adopted a series of measures from this package and decided to submit it for approval to their respective governments. So far, conflicts over larger arms control issues, particularly in respect to nuclear disarmament, between Israel on the one hand and Egypt, Syria and Saudi Arabia on the other have prevented the package from being put into effect. This seems to suggest that there still remains considerable resistance towards the adoption of basic confidence-building measures, let alone a common understanding of security and stability. Therefore even if 'there may not be any country in the region with both the power and intention to threaten Turkey with a major aggression' Turkey's vital interests will remain affected by the lack of security and stability in the Middle East.[93]

Turkey faces also security challenges falling into the second group. Turkey faced mass influxes of refugees, from Bulgaria in 1989 and from Iraq in 1988 and 1991. The danger of another such influx coming from northern Iraq is for the time being deterred by Operation Provide Comfort.

Turkey is also surrounded by obsolete and poorly maintained nuclear power plants. The ones in Armenia and Bulgaria are very close to the Turkish border and constitute a major threat of radiation to the environment. Turkey was among the countries who were badly affected by the radiation from the Chernobyl accident in 1986.

Drug trafficking, and especially the close link between drug trafficking and the PKK, is also becoming a growing security problem.

The growing prospects that Caspian and Central Asian oil will be shipped through the Turkish Straits by super tankers is becoming a major source of concern for the government as well as the inhabitants, particularly of Istanbul. The Bosphorus Strait is characterised by narrow bends and treacherous currents and there have already been numerous accidents involving large ships with considerable damage to the environment.[94] A newspaper report based on a scenario depicting an important section of Istanbul catching fire as a result of two super tankers colliding and turning the Bosphorus into an oil river in flames is probably not too far from reality.[95] This explains why the Turkish Minister of Foreign Affairs commenting on the project to carry Kazakh oil through the Bosphorus argued 'No agreement is worth endangering the lives of 10 million people in Istanbul'.[96]

In an effort to reduce the likelihood of a environmental disaster, the Turkish government introduced the Maritime Traffic Regulations for the Turkish Straits and Marmara Regions in July 1994. The new regulations introduced stricter practices to govern navigation through the Straits. The effect of these regulations has been to slow down traffic and this has led to Russian objections. The Russian government argues that the introduction of these regulations are politically motivated to prevent the transportation of oil by pipeline to Novorossiysk. Also Russia considers these regulations as an interference with freedom of navigation and hence a violation of the 1936 Montreux Convention. Russia by far is the most frequent user of the Straits and in 1993 of the 20,593 passages by foreign ships 6,322 of them were Russian.[97]

Water is scarce in the Middle East and often a source of conflict. Turkey is frequently referred to as a country that has a relative abundance of water, which could be used to meet some of the demand for water in the Middle East.[98] A number of projects have already been put forward to meet this demand. In 1986 the then Prime Minister of Turkey Özal had put forward the idea of a $21 billion 'peace-pipeline' carrying water to Middle Eastern countries. Deep insecurities and the Arab fear of becoming dependent on Turkish water led to the shelving of the scheme.[99] Another project that has attracted headlines is based on the idea of shipping water from the Manavgat river on the Mediterranean coast of Turkey to Israel.[100] In November 1996 during the Turkish Foreign Minister's visit to Jordan, the Jordanians too expressed interest in purchasing water from the Manavgat river.[101] However, finding an economic way of transporting the water still remains the key problem standing in the way of the project.

The idea that water should be treated as a commercial commodity is one that Turkish officials have advocated for some time. Gruen notes that 'Turkish officials have repeatedly stressed that they seek to develop a network of mutually beneficial economic relations among Turkey and all its neighbours. For example, pipelines would convey oil and gas to Turkey and in exchange Turkey would provide water and hydroelectric power to its neighbours...'.[102] *The Economist* has noted that 'the time is coming when water must be treated as a valuable resource, like oil, not a free one like air'.[103] However, this is not an idea that receives much support in the Arab world. Hence, presently, water remains a source of intense conflict between Turkey and two of its neighbours, Iraq and Syria.

The dispute is over the sharing of the water of the Euphrates and the building of the SouthEastern Anatolia Development Project (GAP). This ambitious project began in the 1960s and, when completed, will have 22 dams, 19 hydro-electric power plants and a complex network of irrigation canals. The project is expected to boost Turkey's agricultural and industrial output significantly.[104] The present regimes in Syria and Iraq fear that by controlling the flow of water Turkey might use this project as a political leverage against them. The Turkish government on the other hand argues that it is entitled to exploit water resources within its territories and refutes the idea that Turkey is a water-abundant country. However, the government expresses readiness to share waters equitably as long as neighbouring countries use water in a more efficient and rational manner.[105]

## CONCLUSION

Since the end of the Cold War and the Gulf crisis considerable emphasis has been put on the idea that European security in the broad sense of the word is intimately linked to stability in the Mediterranean region. An early but somewhat unrealistic recognition of this was captured in the proposition that a Conference on Security and Cooperation in the Mediterranean made by the Italian and Spanish delegates during the CSCE meeting in Palma de Mallorca in September 1990. Since then a range of initiatives of varying scopes have been started by the EU, NATO and WEU in addition to other existing fora in the context of the Arab–Israeli peace process. The most ambitious of them all is the one expressed in the Declaration adopted at the end of the Barcelona Conference in November 1995. Accordingly, participants from 12 Mediterranean countries and all 15 EU members accepted 'to establish comprehensive partnership among the participants – the Euro-Mediterranean partnership – through strengthened political dialogue on a regular basis, the development of economic and financial cooperation and greater emphasis on the social, cultural and human dimension, these being the three aspects of the Euro-Mediterranean partnership'.[106]

In respect to building such a partnership, Turkey is in a particularly curious situation. Turkey straddles a number of geographical and political regions, one of which is the Mediterranean. However, presently Turkish officials do not view the world through Mediterranean lenses. Turkish economic, foreign policy and security agendas are dominated by issues perceived to be emanating from other regions. Some of these issues overlap with issues that dominate the concerns of Mediterranean countries while others do not. The economic root causes of terrorism and Islamic fundamentalism as well as immigration, though only to some extent, can be considered as common concerns. As Karaosmanoglu notes at the western end of the Mediterranean many of the problems that threaten stability and security are essentially the product of a typical north–south divide.[107] While the ones in the Eastern Mediterranean additionally have to do with military/ideological issues as well as territorial disputes. Though the threat from ballistic missiles might be considered as a common military challenge to both ends of the Mediterranean, the much heralded two Libyan Scud missiles fired at a base in Lampedusa in 1986 cannot be a match to the potential threats emanating from the arsenal of ballistic missiles and mass destruction weapons in countries neighbouring Turkey.

Nevertheless, some of the most salient issues on Turkey's foreign policy agenda are closely linked to the Mediterranean. Problems relating to Cyprus, Greece and Syria have at least geographically a Mediterranean dimension. However, would a sense of belonging to a common Mediterranean region and institutions help to resolve these problems? It is highly doubtful. At least in the case of the many problems between Greece and Turkey, belonging to common regional institutions has not enhanced the prospects of greater understanding and cooperation between the two countries.

There might nevertheless be a role for Turkey to play in the search for dialogue and stability in the Mediterranean. For decades the Turkish elite has been committed to consolidating a secular and democratic regime. Their task has not been easy and their project is increasingly under threat from political Islam. Political Islam has been gathering force in Turkey as much as in most of the southern Mediterranean countries who are experiencing chronic economic problems. The challenge that awaits those who want to build a stable and secure Mediterranean region have to make sure that the Mediterranean Sea is not perceived as a divide between the Christian 'haves' and the Islamic 'have nots'. It is in this respect that Turkey's role in the Mediterranean might become critical. A Turkey more closely integrated with Europe may well be a Turkey that can help to erode an image of a Mediterranean Sea as a divide between 'haves' and 'have nots' based on religion. At the same time such a Turkey is more likely to reinforce as well as develop its secular and democratic regime. Such a Turkey might be an asset to regional security and stability whether European, Middle Eastern, Caucasian or Mediterranean.

However, a lot is going to depend on developments in Turkey too. The Erbakan government was led by a political party that emphasised Islamic values and symbols in its foreign policy. This was reflected in efforts to develop closer ties with Islamic countries. In the region their preference was to develop relations with radical regimes and groups. Erbakan sought closer relations with Iran and Libya while his party lent public political support to Hamas and Hezbollah at the cost of offending the PLO and Egypt. The Prime Minister's approach to Israel was mostly a function of pressures put on him by his coalition partners, the Turkish Foreign Ministry and the military, rather than his ideological preferences and the will of his party. Clearly, their preference was to reduce the close ties between Israel and Turkey. Under these circumstances it is difficult to see how a government led by the Islamic Welfare Party could have developed a vision of a Mediterranean regional identity that transcended ideological and religious differences.

The junior partner of that coalition government, led by Foreign Minister Çiller, remained committed to the traditional Turkish foreign policy characterised by a preference for enhancing Turkey's ties with Western Europe. She did become much more assertive in her demands from the West but this can partly be attributed to an effort to balance her coalition partner's foreign policy preferences and partly to a need to meet a growing public disappointment with the West's treatment of Turkey. It would be difficult to say that much importance was attributed to the notion of a Mediterranean region by Çiller and her supporters. Instead it is possible to argue that many who supported her foreign policy felt a certain degree of discomfort when Turkey was included among the twelve non-EU countries of the Mediterranean at the Barcelona Summit in November 1995. Similar sentiments were a concern of the Yilmaz coalition government from June 1997.

During the Cold War Turkey was an ally of the West and a general consensus prevailed among the ruling elite about the direction of Turkish foreign policy. Often Turkish foreign policy could be understood as a function of developments at the international level. However, since the early 1990s Turkish foreign policy has become increasingly complex. This is partly due to international developments that have resulted from the end of the Cold War. At the same time it is also partly the product of domestic development in Turkey. Major economic, political and social changes as well as an increasingly assertive political Islam have weakened the consensus on the direction of Turkish foreign policy. A wide range of actors, for reasons ranging from bureaucratic, cultural, economic, ethnic, ideological to religious and political ones, compete to influence foreign policy-making. This competition is going to have an important bearing on which direction Turkish foreign policy takes. Whether it takes an Islamic, a Mediterranean, a Western or a multi-regional direction will, to an important extent, be a product of this competition. Hence, understanding and explaining Turkish foreign policy will increasingly require the need to focus on the domestic sources of foreign policy-making.

## NOTES AND REFERENCES

1. J. Chipman, 'Introduction' in J. Chipman (ed.), *NATO's Southern Allies: Internal and External Challenges*, Routledge, London and New York, 1988, p. 3. For a similar view see also G. Steinberg: 'European Security and the Middle East Peace Process', *Mediterranean Quarterly*, vol.7, no.1, 1996, p. 80.
2. A. M. Williams, 'Turkey: the Mediterranean Context', in C. Balkir and A. M. Williams (eds), *Turkey and Europe*, Pinter, London, 1993, p. 64.
3. M. Clarke, 'Foreign Policy Analysis: A Theoretical Guide', in S. Stavridis and C. Hill (eds), *Domestic Sources of Foreign Policy: West European Reactions to the Falklands Conflict*, Berg, Oxford, 1996, pp. 19–39.
4. V. Hudson, 'Foreign Policy Analysis Yesterday, Today, and Tomorrow', *Mershon International Studies Review*, vol. 39, supp. 2, October 1995.
5. G. Allison, *Essence of Decision: Explaining the Cuban Missile Crisis*, Little, Brown, Boston, 1971, p. 4.
6. G. Özcan, 'Türkiye'de Siyasal Rejim ve Dis Politika', in F. Sönmezoglu (ed.), *Türk Dis Politikasinin Analizi*, Der Yayinlari, Istanbul, 1994, pp. 293–316.
7. B. Tanör, 'Türkiye'de Dis Iliskilerin Iç Hukuk Rejimi' in Sönmezoglu (ed.), op.cit., pp. 317–32.
8. M. B. Aykan, 'The Palestinian Question in Turkish Foreign Policy from the 1950s to the 1990s', *International Journal of Middle East Studies*, vol.25, no.1, 1993.

9. S. Kut, 'Turkish Diplomatic Initiatives for Bosnia-Herzegovina' in G. Özdogan and K. Saybasili (eds), *Balkans: A Mirror of New International Order*, Eren, Istanbul, 1994.
10. Z. Önis, 'Turkey in the Post-Cold War Era: In Search of Identity', *The Middle East Journal*, vol. 49, no.1.
11. G. Winrow, 'Turkey's Relations with the Transcaucasus and Central Asian Republics', *Perceptions*, vol.1, no.1, 1996.
12. A. Y. Saribay, 'Dis Politika Kararlarini Belirlemede Psikolojik Etkenler ve Türkiye', in Sönmezoglu (ed.), op. cit, pp. 383–90.
13. S. Özel, 'Of Not Being a Lone Wolf: Geography, Domestic Plays, and Turkish Foreign Policy in the Middle East', in G. Kemp and J. G. Stein (eds), *Powder Keg in the Middle East*, Rowman Littlefield, Lanham, Maryland, 1995.
14. F. Vali, *Bridge across the Bosporus: The Foreign Policy of Turkey*, Johns Hopkins Press, Baltimore and London, 1971.
15. Clarke, op. cit, pp. 27–8.
16. For Rosenau's discussion of 'theoretical thinking' see J. Rosenau, *The Scientific Study of Foreign Policy*, Frances Pinter, London, 1980, chap. 3.
17. Vali, op. cit, p. 73.
18. *Cumhuriyet*, 8 April 1993.
19. *Turkish Daily News (TDN)*, 14 April 1993.
20. For a first-hand description of the role of the Ministry in respect to these two multi-lateral fora see S. Korkud: 'Nagorno Karabagh Conflict', *Perceptions*, vol.1, no.1, 1996, pp. 146–50.
21. I. Janis, *Victims of Groupthink*, Houghton Mifflin, Boston, 1982.
22. K. MacKenzie, 'Turkey's Circumspect Activism', *The World Today*, vol. 49, no.2, 1993, p. 25.
23. For an analysis of the formation and the role of the NSC in Turkish politics of the 1980s and 1990s see W. Hale, *Turkish Politics and the Military*, Routledge, London, 1994 p. 138 and pp. 290–1.
24. Özcan, op. cit, pp. 302–6; Tanör, op. cit, p. 324 and I. Soysal, *Türk Dis Politikasi Incelemeleri Için Kilavuz (1919–1993), Eren, Istanbul, 1993, p. 37.*
25. Clarke, op. cit, p. 22.
26. R. C. Snyder, H. W. Bruck and B. Sapin (eds) *Foreign Policy Decision-Making*, Free Press, New York, 1963.
27. R. Jervis, *The Logic of Images in International Relations*, Columbia University Press, New York, 1989.
28. M. Kocaoglu, *Uluslararasi Iliskiler Isiginda Ortadogu*, Gnkur. Basimevi, Ankara, 1995.
29. *TDN*, 24 April 1996.

30. *TDN*, 10 May 1995.
31. *Yeni Yüzyil*, 4 May 1995.
32. For an analysis of these results, see A. Çarkoglu, 'The Turkish General Elections of 24 December 1995', *Electoral Studies*, vol.16, no.1, 1997.
33. N. Erbakan, 'Türk Dis Politkasi Nasil Olmali' in *Yeni Türkiye*, vol.1, no.3, 1995, pp. 58–61. See also S. Sayar1: 'Islam and International Relations in Turkey' in S. Mardin (ed.), *Cultural Transitions in the Middle East*, E.J. Brill, 1994; and S. Sayari, 'Turkey's Islamist Challenge', *Middle East Quarterly*, vol.3, no.3, 1996.
34. *Yeni Yüzyil*, 21 April 1996.
35. *TDN*, 4 December 1996.
36. For this world view of the founders of the Turkish Republic see B. Lewis, *The Emergence of Modern Turkey*, Oxford University Press, New York, 1968.
37. Vali, op. cit, p. 335. See also A. Eralp: 'Turkey and the European Community in the Changing Post-war International System', in Balkir and Williams (eds), op. cit, p. 25.
38. Eralp, op. cit, p. 26.
39. Calculated from Table 3 in Z. Önis: 'The Political Economy of Export-Oriented Industrialisation in Turkey', in C. Balim et al. (eds), *Turkey: Political, Social and Economic Challenges in the 1990s*, E. J. Brill, Leiden, 1995, p. 119.
40. See Table X.5 in Y. Kepenek and N. Yentürk *Türkiye Ekonomisi*, Remzi Kitapevi, Istanbul, 1994, p. 261.
41. G. C. Rosenthal, *The Mediterranean Basin: Its Political Economy and Changing International Relations*, Butterworth Scientific, London, 1992, p. 15.
42. D. Vardar, 'Türkiye nin Bati Avpura Kurumlarina Girisi ve Kanoyu', in Sönmezoglu (ed.), op. cit, pp. 378–79.
43. Ibid., pp. 379–80.
44. Eralp, op. cit, p. 41.
45. S. Sayari, 'Turkey: The Changing European Security Environment and the Gulf Crisis', *Middle East Journal*, vol. 46, no. 1, 1992, p. 10.
46. J. Brown, 'Turkey and the Persian Gulf Crisis', *Mediterranean Quarterly*, vol. 2, no. 2, 1991, p. 48.
47 . For a popular example of such a portrayal of Turkey see the feature on Turkey entitled 'Star of Islam' in *The Economist*, 14–20 December 1991.
48. *TDN*, 24 July 1995.
49. *TDN*, 30 Oct. 1995 and 1 Nov. 1995.
50. *TDN*, 25 October 1996.

51. Eralp, op. cit, p. 38.
52. I. Lesser, 'Bridge or Barrier? Turkey and the West after the Cold War' in G. Fuller and I. Lesser (eds), *Turkey's New Geopolitics: From the Balkans to Western China*, Westview, Boulder, CA, 1993, p. 99.
53. *Cumhuriyet*, 30 March 1996. Yilmaz became Prime Minister again in June 1997.
54. W. Hale, 'Turkey and the EU: The Customs Union and the Future', *Bogaziçi Journal*, vol. 10, nos. 1–2.
55. *TDN*, 1 February 1997.
56. *TDN*, 11 December 1995.
57. H. Kramer: 'The EU–Turkey Customs Union: Economic Integration amidst Political Turmoil', *Mediterranean Politics*, vol. 1, no. 1 1996.
58. Statistics for Israel, Lebanon, Malta, Georgia, Armenia, Kyrgzstan and Tajikistan were not available.
59. The 'suitcase trade' refers to large numbers of tourists from CIS countries travelling to Turkey to sell goods they bring with them and in return purchase a wide range of products to take back to their respective countries for personal consumption and/or to resell there. According to an official from the Turkish Foreign Economic Relations Board (DEIK) an estimated $1 billion is spent yearly on Russian goods brought into Turkey while an estimated $5 to $7 billion is spent on Turkish goods purchased by 'suitcase traders', *TDN*, 30 June 1996.
60. *Ekonomik Rapor*, ITO Yayin no: 1996–3, Istanbul, p. 136.
61. As of March 1996, the total work in progress in CIS countries including Russia is $3.5 billion as opposed to $2.2 billion for the Middle East. Ibid.
62. Statistics for Israel, Algeria, Morocco, Malta and Central Asian countries were not available. For the Black Sea group only statistics for Bulgaria and Romania were available.
63. For a general analysis of Turkey economic relations with the Middle East see P. Robins, *Turkey and the Middle East*, Pinter, London, 1991, pp. 100–13.
64. *T.C. Petrol Isleri Genel Müdürlügü Dergisi*, no. 39 (1995), p. 60.
65. Turkey and Serbia prepared to exchange ambassadors for the first time since 1992, *Yeni Yüzyil*, 31 May 1996.
66. *Yeni Yüzyil*, 16 April 1996.
67. M. Akgün, 'Turkey and Russia: Burdened by history and myopia' *Private Eye*: *Quarterly International Review of the Turkish Industrialists' and Businessmen's Association*, vol. 1, no. 3.
68. *TDN*, 4 April 1996.
69. Winrow, op. cit, p.130.

70. *TDN*, 21 April 1995.
71. TDN, 9 May 1996.
72. P. Robins, 'Between Sentiment and Self-Interest: Turkey's Policy Towards Azerbaijan and the Central Asian States' in *Middle East Journal*, Vol. 47, no.4, Autumn 1993, pp.593–610.
73. Winrow, op. cit, p. 136.
74. *TDN*, 2 September 1995.
75. K. Kirisçi, 'Turkey and the Kurdish Safe-Haven in Northern Iraq', *Journal of South Asian and Middle Eastern Studies*, vol. 29, no. 3, 1996, pp. 21–39.
76. S. Kut, 'Türkiye dis politikada tarihsel-psikolojik bir esigi asti', *Milliyet*, 9 May 1996, p. 20.
77. *TDN*, 26 September 1996.
78. H. Bagci, 'Türk Yunan barisi için ilk adimi Türkiye atmali' *Yeni Yüzyil*, 2 October 1996.
79. M. Akgün, 'Pangolos'un açiklamalari' *Yeni Yüzyil*, 28 November 1996.
80. Workshop on 'Turkish-Greek Relations: Prospects for Reconciliation', 30 November 1996, Savoy Hotel, Istanbul. The participants included Thanos Veremis and Theodore Couloumbis from ELIAMEP in Athens.
81. E. Mortimer, 'European Security after the Cold War', in *Adelphi Papers*, no.271, Summer 1992, p.13.
82. Assembly of the Western European Union, thirty-eight ordinary session (second part), Document 1341, *Turkey*, Rapporteur Mr Moya, (6 November 1992), p.4.
83. D. Baldwin: 'Security Studies and the End of the Cold War' (review article), *World Politics*, vol. 48, October 1995, p. 128.
84. The official Turkish governmental position is stated in H. Cetin: 'The Security Structures of a Changing Continent: a Turkish View', *NATO Review*, vol. 40, no. 3, (1992).
85. J. Snyder, 'Arms and Security in the Mediterranean Region' in G. Joffe and T. Niblock (eds), *Security Challenges in the Mediterranean Region*, Frank Cass, London, 1996, p. 179.
86. Özel, op. cit, p.182.
87. A. Karaosmanoglu, 'Turkey and NATO in a New Strategic Environment' paper presented at the conference on 'Mediterranean Security in the Post-Cold War Era' Naples, Italy, 27 February–27 March 1995, p. 15.
88. Quoted in Özel op. cit, endnote 39.
89. See maps in S. Gürel: 'Turkey and Greece:a difficult Aegean relationship', in Balkir and Williams (eds), op. cit, pp. 172–3.
90. *The Military Balance 1995/96*, The International Institute for Strategic Studies, London, 1995, p. 25.

91. Calculated from statistics in ibid., p. 39.
92. *TDN*, 5 April 1996.
93. S. Ergüvenç, 'Turkish Security and the Middle East' in *Foreign Policy* (Ankara), vol. 19, no. 3–4, 1995, p. 9.
94. Between 1 May 1982 and 30 April 1994 there were 208 accidents 25 of which caused fires. S. Oguzülgen: 'The Importance of Pilotage Services in the Turkish Straits for the Protection of Life, Property and the Environment', in *Turkish Straits: New Problems, New Solutions*, Foundation for Middle East and Balkan Studies, Istanbul, 1995, Annexes I and II. Also see Ö. F. Gençkaya: 'Sources of Pollution of the Black Sea: Ships', *Foreign Policy/Dìs Politika*, vol. XX, nos. 3–4.
95. *Yeni Yüzyil*, 9 May 1996.
96. *TDN*, 9 May 1996.
97. G. Kut, 'Major Users of the Turkish Straits and Multilateral Treaties on Environment' in *Turkish Straits* op. cit, p.157.
98. J. Kolars, 'Water Resources of the Middle East' in E. Schiller (ed.), *Sustainable Water Resources Management in Arid Countries: Middle East and North Africa*, special issue of the Canadian Journal of Development Studies, 1992, p. 117.
99. G. Gruen, 'Turkey's Potential Contribution to Arab–Israeli Peace', *Turkish Review of Middle East Studies*, no. 7, 1993, p. 198.
100. Ibid., p. 207.
101. *Yeni Yüyil*, 11 November 1996.
102. Gruen, op. cit, p. 211.
103. 'Water: Flowing uphill', *The Economist*, 12 August 1995, p. 38.
104. For the details of the whole project see *The Southeastern Anatolia Project Master Plan Study: Final Master Plan Report*,4 vols, State Planning Organisation, Ankara, April 1989.
105. G Kut, 'Burning Waters: The Hydropolitics of the Euphrates and Tigris', *New Perspectives on Turkey*, no. 9, 1993.
106. *Barcelona Declaration Adopted at the Euro- Mediterranean Conference* (27 and 28 November 1995), p. 3.
107. Karaosmanoglu, op. cit, p. 2.

# 10 Europe and the Middle East Peace Process: Emerging from the Sidelines

Joel Peters

## INTRODUCTION

At the beginning of October 1996 in a statement issued by the Council of Ministers, the European Union expressed its concern at the outbreak of rioting in Gaza and the West Bank, which had erupted following the opening in September by Israel of the Hasmonean tunnel in the old city of Jerusalem and which had led to the deaths of 64 Palestinians and 15 Israeli soldiers. The European Union essentially held the new Likud-led Israeli government as being responsible for the marked deterioration in the Peace Process and for the outbreak of violence which had been 'precipitated by frustration and exasperation at the absence of any real progress in the Peace Process and [the European Union] firmly believes that the absence of such progress is the root cause of the unrest'. The statement called on the parties to return quickly to the negotiating table and reaffirmed that the search for peace in the Middle East was of vital interest to the European Union. 'Accordingly, the European Union is ready to play an active part in efforts to recommence the negotiations, commensurate with its interests in the region, and on the basis of its major contribution to the Peace Process so far'. In response to the growing sense of crisis, 'The Union is currently preparing a Ministerial Troika visit to the Middle East for talks with the main parties in the Peace Process.'[1]

This statement was followed by a flurry of diplomatic activity by European leaders. Towards the end of the month, French President Chirac made a grand, and highly publicised, tour of the Syria, Israel, Gaza and the West Bank,

Egypt, Jordan and Lebanon in turn. Throughout his visit, Chirac made the case for a greater European role in the Arab–Israeli Peace Process. In his speech to the Palestinian Legislative Council in Ramallah, the first head of state to address the Council, Chirac asserted that a true and lasting peace could only be achieved with the establishment of a Palestinian state.

Chirac's trip was quickly followed by a quieter and altogether lower-key visit to the region by the British Foreign Secretary Rifkind. Rifkind nonetheless took the opportunity, whilst visiting the West Bank town of Hebron, to publicly remind the Israeli government that Britain viewed the building of Israeli settlements in Gaza and the West Bank as illegal and that any expansion of those settlements posed a serious obstacle to the pursuit of peace in the region. During his visit to Abu Dhabi, Rifkind went on to launch a British initiative for the creation of a new regional security forum in the Middle East, modelled closely on the experience of the Organization of Cooperation and Security in Europe (OSCE.)[2]

European concern about the decline in the political situation in the Middle East and the virtual stand-still in the Peace Process between Israel and the Palestinians following the election of Netanyahu as Israel's Prime Minister in May 1996 was reflected in the speech by the Irish foreign minister Spring, in his capacity as President of the European Council, to the opening session of the Middle East and North Africa economic conference held in Cairo in the middle of November. Spring assured the conference that the European Union 'was a friend of the region for the bad times as well as for the good. It is times like these, of a downturn in the peace process, that the Union's support is all the more necessary.' The European Union was committed 'to the peace process for the long haul, until a just and lasting peace can be reached in the Middle East'.

Spring's words were warmly received by the delegates in the hall, especially when he emphasised that the Peace Process should be based on the principle of land for peace and insisted that the economic hardships inflicted on the Palestinians as a result of the prolonged closure of the occupied territories should be relieved without further delay.

This heightened level of diplomatic activity towards the end of 1996 reflected the long-standing frustration within many European capitals at their marginalisation from the central political developments within the Peace Process and a growing dissatisfaction with the American and Israeli self-imposed division of labour; namely that the United States would act as the sole external mediator between the protagonists, whilst Europe would be responsible for providing the bulk of the financial resources required for underwriting any agreements reached by the parties. Europe's growing impatience at being constantly sidelined and being treated as second-string player in the Peace Process was palpable when none of its representatives were invited to attend the emergency Washington summit (convened by President Clinton in the immediate aftermath of the rioting in September to prevent the crisis between Israel and the Palestinians from further escalating). This in spite of the fact the European Union contributes approximately 85 per cent of the running costs of the Palestinian Authority and together with non-EU states such as Norway and Switzerland, the EU's bilateral and multilateral assistance amounts to nearly half of the $2.4 billion pledged to the rebuilding of the Palestinian economy.[3]

At the Special meeting of the European Council held in Dublin on 5 October 1996, European leaders' impatience was palpable. Feeling that they could no longer stand by and watch the Middle East slide into a new cycle of conflict and that the time had arrived to take a more pro-active approach, they decided to appoint a special European envoy to the peace process. Spring informed the economic conference in Cairo that '[t]he European Union is deeply concerned by the recent tragic events in the region as well as by the slow progress of in the direct talks between the parties'. He went on to say that '[the EU] has a responsibility both to the region and to itself to put the Peace Process back on track'. The appointment of the Spanish ambassador to Israel, Miguel Angel Moratinos, as its special envoy to the Peace Process signalled Europe's determination to breathe new life into the moribund peace process and its intent that its future involvement would be more in line with its long-standing and significant political and economic engagement in the region.[4]

## THE VENICE DECLARATION AND BEYOND

Europe's attempt to carve out for itself a separate and distinct role in the quest for peace in the Middle East is not new. Europe has always prided itself on its 'special relationship' with the Middle East and in particular with the Arab world. Europeans, both individually and collectively, have long maintained that they possess a better understanding of the region and therefore a more sophisticated diplomacy than the Americans. Europe's historical legacy, its geographical proximity to the region and its extensive network of political, economic and cultural ties affords it a unique role in, if not responsibility to, the Middle East. As such it is uniquely placed to bring about peace in the region. That this feeling of special affinity is frequently not reciprocated by the Middle Eastern states themselves has not deterred the Europeans states. Nor have the limitations inherent in formulating, let alone implementing, a common foreign policy or Europe's failure to bring about peace in the former Yugoslavia, diminished its belief that it can, or its desire to, play a critical role in bringing about peace between Israel and the Arabs.[5] Although there are differences in thought within the European Union as how best to proceed in playing a more active role in the Peace Process, the policies of the member states do converge around a number of shared principles. Attempts by the member states of the European Union to play a collective role, independent of the superpowers, in the Arab–Israeli conflict can be traced back to the early 1970s.[6]

However, it was the Venice Declaration of June 1980, issued one year after the signing of the peace treaty between Israel and Egypt, which marked the emergence of a distinct and common European stance towards the Arab–Israeli conflict and outlined a collective position on the steps to be taken for its peaceful resolution. Sixteen years on, the Venice Declaration still constitutes the basic principles of European policy towards the Peace Process.

In the preamble to the Declaration the members of the EC stated that 'the traditional ties and common interests which link Europe to the Middle East' obligated them to 'play a "special role" in the pursuit of regional peace'. The Declaration asserted that it was imperative that a just solution be found for the Palestinian problem, and that this issue was 'not simply a refugee one', and that the Palestinian people be allowed to 'exercise fully its right to self-determination'. Significantly, the member states called for the inclusion of the Palestine Liberation Organization (PLO) in any negotiations for a settlement. The Declaration stressed that Israel needed to end its territorial occupation and that the EC was 'deeply convinced that Israeli settlements constitute a serious obstacle to the peace process' and that these settlements 'are illegal under international law'. The Declaration was equally forthright concerning the future of Jerusalem. 'The Nine stress that they will not accept any unilateral initiative designed to change the status of Jerusalem', a stance that the European Union vigorously reaffirmed in its statement following the opening by Israel of the Hasmonean tunnel in September 1996[7].

Israel's response was furious and uncompromising. Two days after the Venice Declaration, the Israeli cabinet issued a statement: 'Nothing will remain of the Venice decision but a bitter memory. The decision calls on us and other nations to bring into the peace process that Arab SS which calls itself "the Palestinian Liberation Organization";... all men of goodwill in Europe, all men who revere liberty, will see this document as another Munich-like capitulation to totalitarian blackmail and a spur to all those seeking to undermine the Camp David Accords and derail the peace process in the Middle East.'[8]

The Venice Declaration marked a low-point in Israel's relations with the European Community from which it has never fully recovered. From the issuing of the Declaration in June 1980 to the convening of the Madrid peace conference in November 1991, Israel vigorously opposed any European attempt to play a significant role in the peace process. In January 1989, Shamir told the chairman of the European Parliament, Lord Plumb, that because of the Community's pro-Palestinian bias it was difficult to conceive of the Europeans as participants in the political process in the Middle East.[9] Israelis from across the political spectrum were united in their criticism of Europe's approach to the conflict, which they saw as representing solely the Arab position. In particular, they were angered by the series of declarations and statements issued by the European Community which, following Israel's invasion of Lebanon in June 1982 and the start of the Palestinian Uprising (Intifada) at the end of 1987, became increasingly critical of Israel's policies and more open in their endorsement of the PLO as the representatives of the Palestinian people and of the rights of the Palestinians to national self-determination. Israel denounced these declarations for their one-sidedness and for the way in which Europe was making increasing demands upon Israel without making equal demands or expecting reciprocal moves from the Arab side.[10] Not surprisingly, the Arab states drew great satisfaction from the stance taken by the European states and saw them as an important counterweight to the near blanket support given to Israel by the United States. Accordingly, they were eager to afford the European Community a greater role in any efforts to resolve the Arab–Israeli conflict.

The positions and diplomacy adopted by the European Community throughout the 1980s did little to advance its ambitions of playing a significant role in bringing about a peaceful resolution to the Arab–Israeli conflict. By mirroring the stance taken by the Arabs, it effectively removed itself as a potential mediator between the two sides. Whilst the same charge, in reverse, could equally well be levelled against the United States, the Europeans had little leverage with Israel and their collective voice was studiously ignored in Jerusalem. Unlike the United States, the European states, at that time, possessed neither the capacity nor any decisive influence over the parties and bring them to the negotiating table. With the end of the Gulf War in the summer of 1991, international efforts to relaunch the Arab–Israeli peace process were intensified. Given the re-invigorated role that the Security Council and the United Nations in general had played in lending legitimacy to the military actions undertaken by the Allied forces to remove Saddam Hussein from Kuwait, there were high hopes within European quarters that an international conference on the Middle East process, under the auspices of the UN, would be convened and that the European Community would play a key role in the next phase of the peace process.[11]

These hopes proved to be short-lived. Neither the Security Council nor the United Nations at large was to play any role in the political developments engendered by the Gulf War. The United States alone took it upon itself to set up an institutional framework to deal with the Arab–Israeli conflict and effectively blocked any alternative international efforts. Whilst Madrid played host, and gave its name, to the conference convened by the United States at the end of October 1991, the European Community was offered only a minor role in the proceedings. Indeed, the United States turned to Moscow, rather than Europe, to act as co-sponsor to the Madrid Conference, despite the fact that the Soviet Union's power was visibly declining and that it was on the verge of collapse. The European Community played no part in the bilateral negotiations that followed the Madrid Conference. Instead, it was invited to participate only in the multilateral talks which were set up by the meeting in Madrid.

In the end, the framework devised by the United States proved to be insufficient to bring about a breakthrough in the negotiations between Israel and the Palestinians. It took a secret back-channel, under the gentle guidance of the Norwegians, for Israel and the Palestine Liberation Organisation (PLO) to put aside their differences and sit down together. European states could draw a sense of satisfaction that a policy they had been advocating since 1980, namely of the necessity of involving the PLO in any negotiations, had finally been adopted by Israel. In reality, however, the Venice Declaration and the strategies subsequently adopted by the Community had little impact on the Arab–Israeli conflict or in bringing Israel and the PLO together. Whilst the European states might have drawn the right conclusions thirteen years previously, they had been less successful in translating their declarations and statements on the Middle East into any effective operative strategy. When Israel and the PLO did turn to a third party, it was to a country, Norway, which had developed good relations with both sides and therefore was able to act as an honest broker. Rightly or wrongly, prior to the signing of the Oslo Accords, the European Community was perceived by Israel as part of the problem and not as part of the solution.

## EUROPE AND THE MULTILATERAL ARAB–ISRAELI TALKS

Denied a seat in the bilateral negotiations, held in Washington under the auspices of the United States, the members of the European Union were consigned to operating solely within the framework of five working groups of the multilateral talks set up by the Madrid conference.[12] The multilateral talks were to run in parallel with the bilateral negotiations. The aim was to bring together Israel, its immediate Arab neighbours and the wider circle of Arab states in the Maghreb and the Gulf to address issues of regional and mutual concern. While the bilateral talks were to concentrate on the political issues of territorial control and sovereignty, border demarcations, security arrangements and the political rights of the Palestinians, the multilaterals would examine a range of economic, social and environmental issues which extend across national boundaries, the resolution of which is a prerequisite for long-term regional development and peace in the Middle East.

The idea of the multilateral track is grounded in a functionalist, liberalist conception of international cooperation and peace according to which the enmeshing of the states in the region in an ever-widening web of economic, technical and welfare interdependencies would force them to set aside their political and/or ideological rivalries. The process of continuing cooperation in areas of mutual concern would blur long-held animosities and would create a new perception of shared needs. Continuous interaction would be accompanied by a learning process which would foster a fundamental change in attitudes and lead to a convergence of expectations and the institutionalisation of norms of behaviour. From progress in the multilaterals would emerge a vision of what real peace might entail and the benefits that would accrue to all parties, thereby facilitating progress in the bilateral talks. Regional cooperation was seen as a powerful tool for stimulating economic growth and social development, for helping reduce regional economic disparities and ultimately for making peace irreversible. Drawing parallels from the experience in Europe, functional cooperation would eventually spill over into regional peace.

That said, the addition of the multilateral track was driven by practical considerations. While the principal purpose was to bring together the regional parties, a secondary consideration was to draw the international community into the peace process. The securing of a lasting and comprehensive settlement of the Arab–Israeli conflict would require marshalling the support, expertise and, above all, the financial resources of the international community.

One of the hopes implicit in the logic of the multilateral track was that it would send a powerful signal that all parties, regional and non-regional alike were committed to the ending of the Arab–Israeli conflict and to the building of a new set of cooperative relations. The widening of the Peace Process in this way was dismissed by many as simply a way of placating Israel, thereby ensuring its participation in the Madrid conference and as a poor consolation prize to the Europeans, and the remainder of the international community, who had been excluded from the sponsorship of the Madrid conference and the substantive issues under discussion in the bilateral negotiations. Bringing together Israel and the Arab states to discuss areas of future cooperation before the core political issues at the heart of the conflict had been resolved was seen as not only idealistic but naive. Few had any real expectations of what tangible benefits might emerge from these talks. However, with the breakthroughs achieved at the bilateral level between Israel and the Palestinians and the signing of the peace treaty between Israel and Jordan, the regional and economic component of the Peace Process gained an increased significance and, with it, an enhanced role for Europe within that process.

The European Union was entrusted with the running of the Regional Economic Development Working Group (REDWG) the largest of the five working groups and the one which reflected most fully the broader goals of the multilateral track. As the 'Gavel-holder' for this working group the European Union has been active in promoting ideas and ventures for future economic cooperation among the parties of the region. During the first three rounds of these talks, held in Brussels (May 1992), Paris (October 1992) and Rome (May 1993) a list of ten spheres of activity was drawn up, and 'shepherds' were assigned to take responsibility for the running of projects in each of these areas. The majority of these projects focus either on infrastructural development or on exploring areas of sectoral coordination. The areas are:

- communications and transport (led by France);
- energy (EU);
- tourism (Japan);
- agriculture (Spain);
- financial markets (United Kingdom);
- trade (Germany);
- training (United States);
- networks (EU);
- institutions, sectors and principles (Egypt);
- bibliography (Canada).

The fourth round of talks took place in Copenhagen shortly after the signing of the Declaration of Principles between Israel and the Palestinians. In the light of the breakthrough at the bilateral level there was an overwhelming recognition in Copenhagen of the need to intensify the workings of the REDWG to ensure that its activities would not become marginalised. Accordingly the group adopted the Copenhagen Action Plan, which outlined thirty-three different ventures. This plan has subsequently formed the working basis of the activities of the REDWG. At the subsequent plenary meetings in Rabat (June 1994) and Bonn (January 1995), the countries responsible have reported on the various activities undertaken and have announced new initiatives within their respective spheres of the Copenhagen Action Plan. In order to finance these activities, the European Union announced at the meeting in Rome that it would allocate $6 million for the preparation of feasibility studies and a further $9.2 million was made available for the preparation of studies and the running of inter-sessional activities to ensure the rapid implementation of the Copenhagen Action Plan.

The European Union has taken the lead in encouraging the regional parties to explore ideas about the future long-term nature of their economic relations and to develop a vision of potential institutional mechanisms and frameworks to support and sustain their efforts towards regional cooperation. As a preliminary step in this direction the European Union convened an informal session at the end of the Rome meeting to allow regional parties to air their views. At the plenary meeting in Rabat, the regional parties agreed upon a number of guidelines and principles to steer their work in the future. Specifically, they recognized the need for:

(1) the pooling of common capacities and the joint tackling of common problems through coordinated efforts;
(2) the removing of obstacles to a more prominent role for the private sector;
(3) the promotion of regional trade, the facilitating of investment and the development of infrastructure;
(4) the encouragement of the free flow of people, goods, services, capital and information within the region.

The working group also agreed in Rabat to establish a smaller monitoring committee which would be staffed by parties from the region. The aim in setting up this committee was to allow the core regional parties – Egypt, Israel, Jordan and the Palestinians – to take a more direct role in implementing the Copenhagen Action Plan, in organising the various sectoral activities and in developing a set of priorities and identifying future projects for the working group. It was also agreed that the committee would be jointly chaired by the core regional participants and by the European Union. The regional parties' co-chairmanship would be held for six months and would rotate alphabetically, with Egypt assuming the chair on 1 January 1995. The specific work of the monitoring committee was divided among four sectoral committees whose membership has been confined to the four core regional parties. The chairing of these committees has been shared out, with Egypt taking on responsibility for work on finance and Israel for trade; Jordan is in charge of promoting regional infrastructure, while the Palestinians chair the committee on tourism.

Having established the Monitoring Committee, it was quickly recognised that a new secretariat was needed to help manage such a broad and extensive programme of activities and to be effective it needed to be located in the region. It was also decided to appoint an executive secretary, provided by the European Union, and staffed by personnel from the region, to service the work of the monitoring committee and the four sectoral committees. Reaching agreement among the regional parties on the permanent location of the secretariat proved to be much harder, and not until the Amman Economic Summit in November 1995 did they all agree that it would be based permanently in Amman. In practice, however, the secretariat, with a skeleton staff had been operating out of Amman since March 1995 and over the past two years has serviced nearly one hundred regional meetings and workshops focussing on practical programmes, often of a technical nature, aimed at promoting regional economic cooperation. Although embryonic in its nature and functioning, the secretariat reflects the tentative steps towards the fashioning of new common structures of cooperation, coordination and decision-making in the Middle East. At this moment at least, it is the first functioning regional institution, spawned by the Middle East peace process and sired by the European Union, in which Egyptian, Israeli, Jordanian and Palestinian officials work together on a daily basis.[13]

Although the primary focus of REDWG has been to foster cooperation at the regional level, it has also proved to be an important forum, especially during the early meetings, for addressing the immediate economic needs of the Palestinians. At the Paris meeting in October 1992, in response to an initiative by the European Community, the World Bank was asked to produce a report on the economy of the occupied territories and to draw up a list of priority projects to overcome the infrastructural restraints to cooperation in the region. After a year of intensive consultations involving several field visits, the World Bank submitted its findings to the meeting in Copenhagen. It is this report which forms the basis for directing $2.4 billion of the financial aid to the Palestinians pledged by the donors' conference held in Washington on 1 October 1993 shortly after the signing of the Declaration of Principles between Israel and the PLO.

The breakthrough between Israel and the Palestinians led not only to a boost in the activities within the multilateral framework but has also paved the way for a more tangible role for the European Union in underwriting the Peace Process and in shaping the new regional economic order. The European Union has emerged not merely as an interested bystander in the welfare of the Palestinians, but as the mainstay of the international aid effort in support of the Palestinian economy. On the eve of the signing of the Declaration of Principles the European Community announced that it would be releasing an immediate aid package of thirty-five million ECUs to enable the PLO to establish minimal services and be in a position to attend to the most urgent needs of the Palestinians. At the donor's conference held on 1 October 1993, the member-states of the European Union collectively pledged an additional 500 million ECUs, spread over a period of five years, for the economic recovery and the developmental needs of the Palestinian territories. This aid package amounted to nearly a quarter of the total funds pledged by the participants at the Washington Conference and made the European Union the foremost donor to the Palestinians.

The European Union also provided $24 million for logistical preparations and was responsible for overseeing the international monitoring operation for elections to the new Palestinian Legislative Council, held at the beginning of January 1996. The European Union has continued to serve as a major contributor to UNRWA and has committed 679 million ECU over the past ten years to the running costs of that organization.[14] To underline their commitment to the long-term development of the Palestinian economy, the European Commission and the Palestinian Authority initialled on 10 December 1996 a Euro-Mediterranean Interim Association Agreement on Trade and Cooperation. This agreement is similar to other association agreements signed with countries in the Mediterranean region, but has been adapted to take into account the special status of the Palestinian territories. The agreement reaffirms the trade concessions which the European Union has offered to the Palestinians since 1986 and looks forward to the development of free-trade over the next five years. The agreement also provides for a wide range of economic and financial cooperation in support of Palestinian industries.

Through its economic commitment to the Palestinians, Europe has acquired a direct and material interest in ensuring that progress is maintained in the negotiations between Israel and the PLO. This led Europe to voice its concern at the breakdown in negotiations between Israel and the Palestinians following the election of Netanyahu and at the marked deterioration of the Palestinian economy arising from the prolonged closure of Gaza and the West Bank, which has prevented Palestinian workers from entering Israel.[15] In response, the European Union put its money where its mouth was, allocating, on 2 October 1996, a further 20 million ECU to alleviate the budgetary problems facing the Palestinian Authority. At the same time, the member states of the European Union have not hidden their growing frustration at being denied a more active political role in the Peace Process. This all the more so considering their financial aid has been critical in covering the immediate short-fall in the budget and in meeting the day-to-day running costs of the Palestinian Authority arising from the security policies and the economic restrictions imposed by Israel, rather than contributing towards the future welfare and long-term development of the Palestinian territories.

## WHAT ROLE FOR EUROPE?

Most European countries have confined their frustrations at being sidelined from the political issues at the heart of the Peace Process to grumbling within the corridors of Brussels. The exception has been France which, under the Presidency of Chirac, has sought to exert a new pro-active role, independent of the United States, for itself and Europe. Shortly after taking office on 17 May 1995, President Chirac made it clear that Europe, as the principal contributor of aid to the Middle East, should be given a greater political say in the affairs of the region. Chirac wasted little time in stamping his own personal imprint on French policy in the region. At the Sharm al Shaykh anti-terrorism conference held in March 1996, he emphasised the importance of meeting the needs of the Syrians and Lebanese in furthering the Peace Process. This was followed up by visits in April to Beirut and Cairo where Chirac expanded on his ideas for an enhanced role in the region for France specifically and Europe in general. In Beirut, Chirac laid out a new initiative for the eventual withdrawal of Syrian forces from Lebanon following the end of Israel's occupation of its self-proclaimed security zone in the south of the country, adding that he would be prepared to send French troops to guarantee any agreement. French activism in the region was taken a step further with the part it played in securing a cease-fire in Lebanon in April following Israel's bombardment of the country, dubbed 'Operation Grapes of Wrath' in retaliation for attacks by Hezbollah on Northern Israel. To the barely disguised irritation of the Americans and the clear annoyance of Israel, French foreign minister de Charette shuttled between Jerusalem, Beirut and Damascus to help broker a deal. Whilst the Americans tried to downplay France's overall contribution, the eventual agreement included the formation of a committee to monitor the cease-fire in which France serves as co- chair along with the United States. France was also instrumental in the creation of the Consultative Group for Lebanon, which is responsible for channelling international aid efforts towards the reconstruction of the Lebanese economy.[16]

Whilst France emerged with its reputation enhanced by its swift response to the crisis in Lebanon, unease was expressed in many quarters at the absence of any consultation and coordination with its European partners. De Charette was immediately shadowed in the region by Italy's foreign minister, Agnelli, accompanied by the other two representatives the EU's troika, making a mockery of European pretensions to speak with one voice. Encouraged by the diplomatic achievements of the spring, Chirac embarked on a high profile tour of the region visiting Syria, Israel, Gaza and the West Bank, Egypt, Jordan and Lebanon. Throughout his trip, he did not refrain from voicing his reservations and the perceived shortcomings of Israel's policies. In his speech at the Techion University in Haifa, Chirac called for the creation of a Palestinian state and for the total withdrawal of Israel from the Golan Heights and Lebanon. He also suggested that the European Union should be included with the United States and Russia as a co-sponsor of the peace process and that the increased involvement of France and the European Union would help restore confidence in the process. Chirac's forthright style was warmly welcomed by the Palestinians and the broader Arab world, particularly after he was involved in a public scuffle with Israeli security guards whilst touring the Old City of Jerusalem. Not surprisingly Chirac's pronouncements received a frosty reception in Jerusalem, reviving deeply held beliefs in Israel that Europe remains strongly biased towards the Arab world.[17]

Chirac's tour won him acclaim within the Arab world. In Europe, however, criticism was voiced at his style of his diplomacy, which many felt was aimed more at promoting French interests in the region – not to mention his own prestige – rather than advancing Europe's role in the peace process. British Foreign Secretary Rifkind rushed to the region. Whilst his message was little different to Chirac's, the tone of the visit was fundamentally different, revealing that although Europe might be of one mind over the future of the peace process it did not necessarily speak with one voice. France has come under attack for embarking upon independent initiatives in the Middle East and for failing to coordinate its activities with its European partners. At the same time, however, it has led to increased diplomatic activism in the region by the European Union as a whole and has spurred a broader, and much overdue debate, about Europe's potential role in the Peace Process.[18]

One of the immediate outcomes of Europe's determination to play a more visible political role was the decision taken in Luxembourg at the end of October 1996 by the Ministers meeting of the General Affairs Council, to appoint a special European envoy to the peace process. As noted above, this position was filled by Moratinos, a long-standing and well respected diplomat, who at that point was serving as the Spanish ambassador to Israel. Whilst the decision to appoint a special envoy to the region received much attention, there was less certainty as to the role and specific functions that he would perform. Moratinos' formal mandate was as broad as it was vague:

- to establish and maintain close contact with all the parties to the peace process, other countries of the region, the United States and other interested countries;
- to observe peace negotiations between the parties, and to be ready to offer the European Union's advice and good offices should the parties request;
- to contribute where requested to the implementation of international agreements reached between the parties, and to engage with them diplomatically in the event of non-compliance with the terms of these agreements;

- to engage constructively with the signatories to agreements within the framework of the Peace Process in order to promote compliance with the basic norms of democracy, including respect for human rights and the rule of law;
- to monitor actions by either side which might prejudice the outcome of permanent status negotiations.

It was also unclear which bodies the new envoy would be directly responsible to and how his presence in the region would best complement the existing institutional frameworks for the implementation of European policy. It was decided that the special envoy would be guided by, and report under the authority of the Presidency to the General Affairs Council and that he would report to the Council's bodies, 'on possibilities for EU interventions in the peace process, and on the best way of pursuing EU initiatives and ongoing [peace-process] EU business including the political aspects of relevant EU development projects'. At the same time, it was also stressed that the envoy would carry out his tasks 'without prejudice to the role of the Commission' which would be fully associated with his activities.[19]

In the wake of President Chirac's visit and the Council of Ministers' Declaration of 1 October following the opening of the Hasmonean tunnel, the decision to appoint a special European envoy to the Peace Process was not greeted with great enthusiasm by Israel. The Israeli government reacted to the 1 October Declaration in angry tones reminiscent of its denunciation of the Venice Declaration and other European statements of the 1980s. Efforts for peace in the Middle East would not be helped 'by a partial European position or by language that could have a threatening tone... Europe's desire to contribute to the peace process should not be manifested by adopting positions that could hinder the peace process and the negotiations aimed to promote it'.[20]

The appointment of the special envoy was seen by Israel as yet a further sign of renewed European efforts to meddle in the Peace Process and to curry favour with the Arabs by exerting public pressure on Israel, though it did quietly welcome the appointment of Moratinos to that post instead of the high-profile political appointee as had originally been envisaged by some European states.

Given the vagueness of his mandate and the initial lukewarm reception to his appointment by Israel, it was unclear what specific functions Moratinos might perform in furthering European interests in the peace process. However, following his appointment at the end of October 1996, Moratinos quietly carved out a role for himself. Attracting at first little public attention, he set out by familarizing himself with the main actors of the region and by gaining their confidence. Careful not to tread on the toes of the Americans, Moratinos remained firmly in the background during protracted efforts by the US special coordinator Dennis Ross to mediate an agreement between Israel and the Palestinians for the withdrawal of Israeli troops from Hebron. At the time of the agreement, considerable attention was focussed on the content of the letters of assurance which US Secretary of State Christopher had sent to both sides. These letters were seen to provide an important sense of surety, finally allowing Arafat and Netanyahu to put their signatures to the agreement. Unknown at the time, Moratinos was working quietly behind the scenes and the European Union too supplied Arafat with a letter assuring him that it would 'use all its political and moral weight to ensure that the agreement would be 'fully implemented.'[21].

The presence and the meditation of Moratinos, representing the political and economic weight of the European Union in the region, affords all the parties an additional diplomatic environment in which to work towards resolving their differences. This diplomatic space offered by Europe can bear fruit in the mediation efforts aimed at finding a formula to bridge the positions of Syria and Israel in order to bring them back to the negotiating table. It is notable in this respect that after the signing of the Hebron deal, Moratinos directed his attention to reviving the Israeli–Syrian negotiations and devoted considerable time to shuttling between Jerusalem and Damascus (as well as Washington) to this end.

Europe also has the capacity to play a critical role in fostering the conditions for greater regional cooperation between the states of the region, especially in the economic sphere. The Arab–Israeli multilateral talks are not the only context in which Europe has been engaged in trying to develop new structures of cooperative relationships in the region. In November 1995, the European Union launched in Barcelona its Euro-Mediterranean partnership – the Barcelona process – aimed a developing a new framework of peaceful and cooperative relations in the Mediterranean region. To this end, the European Union committed 4.7 billion ECU over the following five years for infrastructural development and financial adjustment for the projected creation of free-trade zone across the Mediterranean region by the year 2010. The Barcelona process envisaged not just the eventual integration of the economies but also the development of new cooperative frameworks encompassing future political, security and civil relations between the states of the Mediterranean region. The development and the launching of the Barcelona process received its impetus from the progress achieved in the Arab–Israeli Peace Process. However, the Barcelona process should not be viewed as a competitive framework to the Peace Process, but rather as providing a complementary diplomatic environment to help reduce tensions between Israel and the Arab states. Indeed, one of the Euro-Mediterranean partnership's early achievements was to bring Syria and Lebanon to the same table as Israel in a multilateral forum, something which the multilateral peace talks and the Middle East and North African Economic Summits had failed to achieve.[22]

Furthermore, whilst the multilateral talks were effectively suspended following the election of Netanyahu and the stagnation in the Peace Process, Israel and the Arab states continued to sit down together in the programme of follow-up meetings, covering issues such as water resources, industry and energy policy, tourism and environment which took place under the oversight of the Barcelona Euro-Mediterranean Committee. The social and cultural provisions of the Euro-Mediterranean initiative also allowed for the bringing together of Israeli and Arab civil society, an element which is essentially for the long- term reconciliation of these societies and for the development of a lasting and comprehensive peace in the Middle East.[23]

## CONCLUSIONS

European Union has begun to assume a critical and indispensable political role in the Peace Process. The European

Union cannot supplant the primary role of the United States. Nor should it aspire to. Yet, with its capacity to express a greater even- handedness in attending to the political, social and economic needs of the Arab side, Europe can play a valuable role in supplementing the efforts of the United States and help in shaping the future priorities and direction of the peace process. To do this, European leaders will have to construct a more nuanced policy than those developed in the past. They will need to find the balance between pragmatic coordination, and playing second fiddle, to the United States and the launching of new independent initiatives. It will also demand the development of a greater degree of coordination amongst the member-states themselves, whose policies have often been driven by long-standing bilateral rivalries and the promotion of their own commercial and national interests in the region. If the European Union can rise to this challenge and overcome its own internal contradictions then the benefits may well be two-fold. Not only would the European Union play an important part towards the development of a true, comprehensive and lasting peace in Middle East. It would also illustrate the potential, and serve as a small, yet important, step in the construction, of a Common Foreign and Security Policy (CFSP) for the European Union. It remains to be seen whether it can meet this challenge.

## NOTES AND REFERENCES.

1. 'Declaration by the European Union on the Middle East Peace Process', General Affairs Council, 1 October 1996.
2. See Malcolm Rifkind, 'Blueprint for a Region at Peace', *The Times*, 5 November 1996.
3. See Claire Spencer, 'New European Approaches to the Middle East Peace Process?' *Brassey's Defence Yearbook 1997* and 'Europe Spurned', *Middle East International*, 2 October, 1996, p. 6.

4. See *Europe in Israel* (Newsletter of the delegation of the European Commission to the State of Israel), no 37, December 1996, p.1.
5. François d'Alacon, 'The EC looks to the a New Middle East', *Journal of Palestine Studies*, vol. 23, no.2 1994, pp. 42–3.
6. Ilan Greilsammer, 'Failure of the European "Initiatives" in the Middle East', *Jerusalem Quarterly*, 33, Fall 1984, pp. 40–9.
7. For text of the Venice Declaration see Walter Laqueur and Barry Rubin (eds), *The Arab–Israeli Reader: A Documentary History of the Middle East Conflict*, pp. 414–15.
8. Quoted in Rosemary Hollis, 'Israeli–European Economic Relations', *Israel Affairs*, vol. 1, no. 1, 1994, p. 125.
9. Ibid., p. 126.
10. See Ilan Greilsammer, op. cit., p. 47.
11. John Palmer, 'The European Community and Middle East Peace,' *Middle East International*, 16 August 1991, pp. 17–18.
12. The multilateral talks consist of five working groups – arms control and regional security, environment, water, refugees and regional economic development. For details of the five working groups and these talks see Joel Peters, *Pathways to Peace: The Multilateral Arab-Israeli Peace Talks*, Royal Institute of International Affairs, London, 1996.
13. For a detailed analysis of the Regional Economic Development Working Group (REDWG) see Joel Peters, op. cit., pp. 46–60.
14. For details and breakdown of European aid to the Palestinians see Communication from the Commission to Council and Parliament on future European Union economic assistance to the West Bank and Gaza Strip, 19 September 1996. http://europa.eu.int/en/comm/dg1b/en/cisjoren.htm.
15. See Joel Peters, *Israel Under Netanyahu*, Hellenic Foundation for European Foreign Policy, Occasional Papers, January 1997.
16. See Ailie Saunders, 'Chirac and the Middle East', *RUSI Newsbrief*, vol. 16, no. 5, May 1996, pp. 35–37 and 'France and the Middle East', *Middle East Economic Digest*, 25 October 1996, pp. 9–10.
17. See David Makovsky, 'Behind Chirac's Foreign Policy Activism', *Jerusalem Post*, 25 October 1996. Chirac's original itinerary, to the fury of Israel, did not include a visit to the Israeli Knesset.
18. See Claire Spencer, 'New European Approaches to the Middle East Peace Process?' op. cit., and Rosemary Hollis, 'Europe and the Middle East: Power by Stealth?, *International Affairs*, Vol. 73, No. 1, 1997, pp. 15–29.

19. See *Europe in Israel*, op. cit., pp.1–2.
20. Quoted in Rosemary Hollis, op. cit.
21. For text of the letter see *Jerusalem Post*, 10 February 1997.
22. Syria and Lebanon have consistently refused to attend any of the meetings of the multilateral talks, arguing that these meetings were premature and that the Arab states should not discuss regional cooperation with Israel until a political settlement is reached at the bilateral level.
23. See Claire Spencer, op. cit.

# Index

*Note*: page numbers in **bold** type refer to illustrative figures or tables; references are to the Mediterranean area unless otherwise indicated.